Religion, Magic, and Science in Early Modern Europe and America

Religion, Magic, and Science in Early Modern Europe and America

Allison P. Coudert

Praeger Series on the Early Modern World
Raymond B. Waddington, Series Editor

PRAEGER

AN IMPRINT OF ABC-CLIO, LLC
Santa Barbara, California • Denver, Colorado • Oxford, England

Library of Congress Cataloging-in-Publication Data

Coudert, Allison, 1941–

Religion, magic, and science in early modern Europe and America / Allison P. Coudert.

p. cm. — (Praeger series on the early modern world)

Includes bibliographical references (p.) and index.

ISBN 978–0–275–99673–4 (alk. paper) — ISBN 978–0–275–99674–1 (ebook)

1. Religion and science—Western countries—History. 2. Occultism and science—Western countries—History. I. Title.

BL245.C68 2011

261.5509182′109031—dc23 2011034900

ISBN: 978–0–275–99673–4

EISBN: 978–0–275–99674–1

15 14 13 12 11 1 2 3 4 5

This book is also available on the World Wide Web as an eBook.
Visit www.abc-clio.com for details.

Praeger
An Imprint of ABC-CLIO, LLC

ABC-CLIO, LLC
130 Cremona Drive, P.O. Box 1911
Santa Barbara, California 93116-1911

This book is printed on acid-free paper ∞

Manufactured in the United States of America

To Paul A. Castelfranco

“Gutta cavat lapidem non vi, sed saepe cadendo;
sic homo fit sapiens non bis, sed saepe legendo”
Giordano Bruno, *Il Candelaio*

Contents

ACKNOWLEDGMENTS

The early modern period was one of transition, when people were drawn to the past—to the idea of the lost Golden Age and the Garden of Eden—but also to the future, envisioning all kinds of marvels from submarines and flying machines to running water, frozen chickens, blood transfusions, and heated bedrooms with en suite bathrooms. It is with this Janus-faced vision in mind that I look back on my good fortune to have been a graduate student at the Warburg Institute when Ernst Gombrich was its head, and Frances Yates, D. P Walker, and Charles Schmitt were there to guide me over the shoals of Renaissance and early modern intellectual history. It was they who planted the idea in my mind that well into the eighteenth century, religion, magic, and science were all of a piece. This was the premise on which the Warburg Institute had been founded, and it is one that has stuck with me through all the years of my academic life and is at the core of this book.

Remembering my years as a graduate student learning to write ambidextrously so I could sit on and warm at least one hand in the frigid reading room of the Warburg and thinking back to the lunchroom and all those frothy cappuccinos tasting of eggs (because the nozzle of the cappuccino machine steamed milk and scrambled eggs) takes me back to what I now consider my past perfect. There was Frances Yates taking nail scissors from her ample purse to cut through the impossibly thick plastic cover trapping the little piece of cheese that came with crackers for lunch. There was Professor Gombrich and other members of the faculty with their guests, all sitting at the same low tea table with students around them in one of the most democratic and terrifying environments a student could possibly imagine. People, by which I mean, adult academics, actually paused to think before they spoke. Death at the desk was not unexpected, given the age of most of the Warburg regulars, which made it all the more wonderful when a fellow graduate student rushed into the reading room to announce that a couple had been observed kissing in the stacks!

As I look back on my years at the Warburg, I envision a perfect past just like my early modern predecessors. But I am equally Janus-faced and can appreciate the wonders that the Warburg Institute set in store for me as I ventured into the academic life I have thoroughly enjoyed—well, really, loved. Having Frances Yates as my "tutor," as a thesis advisor was called, was an immense stroke of luck. She probably was not the best tutor because she was so interested in everything and never chastised me fiercely enough if I got off track. But in reality that was her gift. Her enthusiasm for uncovering what was unknown or had been hidden or obliterated was what I remember most. Although she hardly went anywhere except in her imagination, her spirit of adventure remains for me a stunning vindication of academic life and a valid refutation of the tired old saw that "those who can do and those who can't teach."

Understanding the past is the only way to understand the present. Historians like Frances Yates were wholly aware of the missed opportunities the past presented only to be lost through human obtuseness, stupidity, and greed. But this did not make her despair as it could have for someone experiencing such dramatic, even traumatic, changes in her own life with the death of her brother in World War I and the tremendous transformations she experienced growing up in the Edwardian era and dying in 1981. None of that stopped Miss Yates, as she was called by most of us, from seeing marvels all around. I remember her remarking when she returned from a lecture tour in the United States that the parking lots in Princeton looked like seas of silver in the drab winter landscape as the sun glinted off car roofs. Hers was a novel way of seeing what most people would dismiss as mundane. It is therefore especially with Frances Yates and the Warburg Institute in mind that I look back to the past in gratitude. But it is Paul and Marie Castelfranco, who generously endowed the Chair I now hold in the History of Christianity in the Religious Studies Program at Davis, whom I have to thank for the stimulating intellectual life have I led for the past six years.

I cannot think of anyone more suited to the dedication of this book than Paul Castelfranco. He and Marie recognized that our modern tendency to compartmentalize life distorts the reality human beings face every day. Religion and science were of a piece in Paul's long career of teaching and research at Davis, and they continue to be in his retirement. Paul and Marie were founders of the Methodist Church in Davis and long-time supporters of the Religious Studies Program. Talking to Paul is to live in the present but to go back to the intellectual world of the Warburg Institute, where scholars seemed to know so much more than we do now, be it ancient languages, history, science, or philosophy.

Were it not for the Castelfrancos I would not be at UC Davis. The years I have spent in the Religious Studies Program watching it grow with the addition of truly exceptional and innovative scholars has been nothing short of exhilarating. I have never been in a happier and more congenial intellectual environment, and for this I thank my colleagues Naomi Janowitz, Keith Watenpaugh, Baki Tezcan, Catherine Chin, Mark Elmore, Archana Venkatesan, Flagg Miller, Moulie Vidas, Wendy Terry, and Meaghan O'Keefe. I would also like to thank colleagues in the History Department at Davis, notably Sally McKee and Daniel Stolzenberg, who have enriched my knowledge in innumerable ways.

Much of this book would not exist had it not been for the many enjoyable hours I spent with Taylor Corse translating Anne Conway and F. M. van Helmont while discussing early modern philosophy and life in general. Diane Wolfthal, another colleague from my years at Arizona State University, made walking the treadmill at the recreation center an intellectual experience. I have also been helped enormously over the years by Professor Albrecht Classen at the University of Arizona, whose erudition, wise counsel, and international symposia provided the seedbed for many of the ideas in this book. My year at the Advanced Institute for Judaic Studies at the University of Pennsylvania, where the Institute's director David Ruderman, together with Moshe Idel and Guy Stromsa, organized a seminar on Christian Hebraism was pivotal. It marked a milestone in my appreciation of the fruitful ways Christians and Jews interacted in early modern Europe and reinforced the importance of esoteric currents of thought like the Kabbalah in Western history. The year I spent at the Institute of Advanced Studies in Princeton in the seminar of early modernists organized by Jonathan Israel provided another invaluable opportunity to think through major themes in this book. I have also benefited greatly from contact and conversations with members of the Association for the Study of Esotericism (ASE), The European Society for the Study of Western Esotericism (ESSWE), and The Group on Western

Esotericism at the American Academy of Religion. I thank Antoine Faivre, Jean-Pierre Brach, Joscelyn Godwin, and Wouter Hanegraaff for welcoming me to into the fold of Esotericists, where I have had the good fortune of meeting many scholars in the field, among whom I am especially indebted to Arthur Versluis, Marsha Keith Schuchard, Kocku von Stuckrad, Thomas Hakl, Marco Pasi, and Cathy Gutierrez. Lastly, I thank my good friends in the Davis community Linda Clark, Joan Hogan, and Janet Regnell for helping me remove some of the more egregious errors in this manuscript. Whatever merits this book may have were greatly enhanced by the patient editorial skills of Ray Waddington, the general editor of this series on the early modern world.

Introduction

It is a daunting task to write a book about not one, not two, but three subjects that according to many fine scholars do not exist, namely religion, magic, and science. Winston L. King is one of many authorities in the field of Religious Studies to conclude that determining what religion is, is "a hopeless task."[1] Having struggled with the problem of defining magic, Olof Pettersson suggests it should be given "a decent burial."[2] And when it comes to science, Bruno Latour simply says, " 'Science'—in quotation marks—does not exist."[3] The response to such a wholesale rejection of the topics of this book cannot be in the same vein as the famous remark made by Justice Potter Stewart when called upon to define pornography, "I know it when I see it." Many of us may think we know religion, magic, and science when we see them, but the truth is we don't, and this book is about why we don't and how what we think we know about all three came into existence during the early modern period itself. Our definitions of religion, magic, and science are just that, ours, modern definitions that have a long and contested history.[4] Words, like ideas, beliefs, and institutions, have histories, and having a history means that things have not always been the same but change with changing circumstances. While this seems obvious, the implications are not always understood, much less accepted. For if language changes and the meanings of words are unstable, where is the Archimedean point from which we can view the world? This is an issue that has plagued authors from ancient times to the present. To put the issue in its simplest historical terms: if language is a gift of God and words consequently reflect a divinely ordained reality, then embedded in language are absolute, immutable meanings that lead to absolute, immutable truths. But if language is a human creation that changes over time, then man, to quote Clifford Geertz's riff on Max Weber, is "an animal suspended in webs of significance he himself has spun."[5] Implicit in these two views of language are diametrically opposed concepts of what humans are and how they should conduct themselves. According to the first view, language embodies God-given truths, and it is incumbent upon men to use it properly and respectfully. This naturally raises the issue of who is to decide what is "proper" and "respectful," but there have been plenty of individuals and institutions willing to decide just this—one has only to the think of the Inquisition, the Catholic Index of Prohibited Books, and the innumerable governmental institutions involved and still involved in censorship. But if language is a human invention, then there is no higher authority that individuals can invoke when they speak. Man is essentially responsible for himself; he is both his own creator and the creator of the world. This does not mean that there is no physical universe out there, although some people like George Berkeley (1685–1753), Irish Bishop and philosopher, argued there isn't. It simply means that the way humans see the world is conditioned by human perception and has no necessary connection to what the universe actually is and looks like. And this brings us back to the issue of "religion," "magic," and "science." These terms are confusing precisely because they have been created and used by human beings for their own purposes. They are no more or less obscure that any other term such as "reason" or "culture" or "God," for that matter. All words and concepts are slippery inasmuch as they can be understood in different ways by different people. This does not mean we should "bury" them and stop talking as did Jonathan Swift's "language projectors" in The Grand Academy of Lagado described in *Gulliver's Travels*. Because language was so imprecise, these men conversed by carrying around sacks of things, which they presented to each in silence.[6]

The postmodernist impasse about the validity of using terms like religion, magic, and science comes at the end of a long debate that became particularly acute during the early modern period when language was a topic of compelling interest. Every major figure, and these include Reuchlin, Rabelais, Paracelsus, Agrippa, Postel, Böhme, Kircher, Hobbes, Descartes, Comenius, Spinoza, Locke, Boyle, Newton, Leibniz, Condillac, and Condorcet, as well as a great many lesser ones, were deeply concerned with language, not in the narrow sense of linguistics but in terms of the wider philosophical, political, and religious issues that lie at the heart of one's conception of what language is and how it functions. As historians have come increasingly to realize, language was a central issue throughout Western history, as it is indeed today. Nancy Struever argues that the history of rhetoric is nothing short of cultural history writ large in terms of two competing models of what language is and how it functions: one, in which language reflects an ideal realm of eternal truths, and a second, originally fashioned by the Greek Sophists, in which language creates the realities it describes.[7] Although Struever's book is devoted to the Renaissance, her thesis applies equally well to the early modern and later periods. Language was and still is a preoccupation because as a means of expression and communication language is what connects people to the world and each other. Any discussion of language therefore inevitably involves an examination of how people think and use words, and what words actually mean. Thus language is an integral part of anthropology, psychology, history, philosophy, science, and religion. The debate about whether language is natural or artificial, ordained by God or created by man, an innate capacity or learned from sense experience is neither neutral nor dispassionate. The answers given to each question entail a web of consequences that could and still can lead to arrest, imprisonment, even execution. Take the following snatch of conversation at the University of Paris in the 1820s reported by Hippolyte Taine:

> "Are you still a sensualist, immoralist and atheist?"
> "Why do you say that?"
> "Well, after all, you deny that reason is an independent faculty. You deny the existence of innate ideas. You maintain that a perfect science is nothing more than a perfect language. You continue Condillac's line of thinking, thus you can believe in neither truth, nor justice, nor God."
> "Great God!"
> "You are basically a man of the eighteenth century. Your philosophy destroys the dignity of humanity. You are either a materialist or a skeptic."[8]

If one substitutes Hobbes for Condillac, the quotation applies equally well to the seventeenth century.

The seventeenth century debate about language was itself the product of earlier developments and speculations. The acrimonious controversies between Catholics and Protestants over the sacraments, miracles, the role of saints, the power of relics, and the effectiveness of such rituals as exorcism and baptism all involved the question of whether words, gestures, and specific substances (holy water, the cross) possessed intrinsic power. Did the wafer and wine actually become Jesus's body and blood at the exact moment the priest said "Hic est corpus meum" and "Hic est enim calix sangunis mei," or were the wafer and wine simply symbols recalling Christ's sacrifice? Had miracles ceased? Could humans invoke angelic and demonic spirits, and could these spirits interact physically with human beings? Who had the right to speak for God? Subsumed in these debates were others about the efficacy and legitimacy of magic and the occult, the existence of witchcraft, and the nature of science and scientific evidence. Language, or, more accurately, attitudes toward language, played a key role in every one of these issues and debates. What emerges with increasing clarity in the early modern period is the unsettling realization that language is not rooted in the world of actual things, that words and things are therefore not one and the same; their relationship is a matter of convention. Even more unsettling was the idea that things themselves are no more "real" than words. Gilbert and Sullivan's tag line, "things are seldom what they seem," catches the reaction of many people to the emerging science of the day. With the invention of telescopes and microscopes it became obvious that the human senses were not adequate instruments of perception. Human senses could not give a picture of the actual world. Science could, but only because it was nonsense!

This detour into the way ideas about language and reality changed during the early modern period is crucial for understanding what was meant by the terms religion, magic, and science at the time. As we shall see, the meaning of all three changed significantly from the sixteenth to the eighteenth centuries. While at the beginning of the period it is virtually impossible to separate one from the other, although many people tried, by the end a separation had been made. The separation and the definitions given tell us more, however, about those who made them than the actual situation. Being "modern" meant that one rejected magic as "primitive" and embraced science as "rational" and "civilized." It also meant that one drew a line separating the human from the non-human, nature from culture, and the natural from the supernatural. The problem was and still is that most people do not really hold to these lines of separation. Magic has not gone the way of the dinosaurs; it still exists, and, for most people the natural world is still "enchanted." Although we are different from people in the early modern world for reasons that will become clear, we are not quite as different as we think.

The early modern period was one of stark contrasts: witch burnings and the brilliant mathematical physics of Isaac Newton; John Locke's plea for tolerance and the palpable lack of it; the richness of intellectual and artistic life and the poverty of material existence for all but a tiny percentage of the population. Whether rich or poor, ruler or ruled, the situation most people found themselves in was one of extreme insecurity. Life expectancy in England during the 1640s was thirty-seven years. Fifty percent of children died before the age of ten. Small pox affected eight out of ten people, killing one in seven and disfiguring the majority of the rest.[9] Plague was desperately and justifiably feared. In 1635, all the inhabitants of Malpas, a small hamlet in Cheshire, died of the infection. The last man to fall ill actually dug his own grave and buried himself.[10] There was no clear idea of contagion or of the importance of sanitation and personal hygiene in combating disease. Samuel Pepys hardly ever washed more than his hands, face, and neck until his wife went to the bathhouse and refused to allow Pepys in bed with her unless he bathed. Pepys stood firm for three days and then relented.[11] The habits of King Charles I and his courtiers were even less refined. Anthony Wood describes the foul state of Christ's College after the court fled from Oxford during the Civil War: "Though they were very neat and gay in their apparel," he wrote, "yet they were very nasty and beastly, leaving their excrements in every corner, in chimneys, studies, cole-houses, cellars."[12] Under such conditions, typhus ran rampant, parasitic worms were a common and debilitating complaint, and infection a prevalent, all-too-often mortal, affliction. Coughs, sores, aches, and itches were a part of life in cold, drafty houses in an age supremely democratic when it came to lice and fleas.[13]

The lamentable lack of sanitation in the early modern period ensured that tainted food and water were a constant source of illness. Milk sold in open pails from infected cows in filthy urban streets was a breeding ground for tuberculosis as well as food poisoning. In the seventeenth and eighteenth centuries tuberculosis was responsible for one-fifth of all the deaths in London, and the disease was particularly severe for those under four.[14] Fly- and worm-ridden meat and adulterated or unsanitary grain and water ensured that dysentery—otherwise known as the "bloody flux," "lusk," "surfeit," or, more graphically, "gripping of the guts"—was a common complaint and explains why every book of physic contains remedies for "stomach ache," "vomiting and diarrhea," "worms," "diarrhoea and dysentery," "Colic pain in the Bowels or Gripes in the Stomach," and "hemoerage of the stomach." Poor sanitation brought with it typhus and a variety of fevers that medical practitioners attempted to cure with remedies that included salves, bandaging, and drinks made with such things as beer, wine, nutmeg, cream of tartar, roasted hare's heart, and cobwebs. Constipation and gout were particular problems for the upper classes with their excessive consumption of meat. Given the lack of milk and butter in most diets, kidney and bladder stones were frequent and excruciatingly painful and rickets common among children. A leitmotif of repulsive ingredients runs through the medical recipes to cure these various ailments: "dog," "swine," or "mule manure," "Powder from rain-worms," "a cloth with which a corpse has been washed," "a human bone from a church yard," "a toad dried and powdered," "mouse excrement," "whitened dog excrement (caused by eating bones)," "a hedgehog or Goats hoofs," "the mucus or phlegm (foam) from a tired horse." One can imagine the reaction of the child subjected to the following medical procedure: "To cure the Thrush, take a living Frog, and hold it in a cloth, that it does not go down into the Child's Mouth, till it is dead; and then take another Frog."[15] It is a comfort to know that chicken soup was an old staple then as it is now: ". . . after delivery [of a child] immediately give chicken or capon broth (killed as soon as mature)."[16]

In addition to the medicaments described above, the medical profession had two other standard therapies to fall back on, blood-letting and purging, but how effective any of these remedies were must be left to the reader to judge. Take the case of the "Sun King," Louis XIV. Unlike ordinary people, this most glamorous of monarchs had any number of barbers, surgeons, and physicians at his beck and call, but whether that proved a blessing or curse is debatable. Admittedly he still hunted at the age of 74, but this was in all likelihood due to his own iron constitution rather than the ministrations of his doctors. He survived gonorrhea, typhus, measles, malaria, renal colic, gastric problems (caused by gluttony), constipation, vertigo, carbuncles, and gout. His staff of medical men treated these various afflictions with the same therapies of blood-letting and enemas—all in all he was "purged" some 2,000 times over the course of his 77 years. In 1685 his surgeon-dentists split his left upper jaw attempting to fix his decaying teeth; the result was that whenever he drank liquid squirted from his nose.[17] It is understandable that the standard work on surgery of the day, Richard Wiseman's *Severall Chirurgicall Treatises* (1676), was popularly known as "Wiseman's Book of Martyrs."[18]

Given the stark realities of what medicine had to offer, it was no wonder that many people cast a wary eye on the profession and in their desperation to find relief from everyday assaults on body, mind, and spirit resorted to clergymen, self-styled prophets, astrologers, wise men, cunning women, charlatans, quacks, mountebanks, and even witches to alleviate their fear, pain, and downright misery. As Richard Burton wrote in 1621, "Sorcerers are too common; cunning men, wizards, and white witches, which, if they be sought unto, will help almost all infirmities of body and mind."[19]

Early modern men and women had more than disease to contend with. Violence was never far from the surface. The sixteenth and seventeenth centuries were marked by continual crisis in both the social and political world. It was a period of vicious religious controversy and non-stop warfare—only seven years out of a hundred saw peace. The concept of toleration was meaningless to the vast majority of people, who were ready to defend their beliefs to the death. Aggression and brutality were common features of daily life. Death could come from the most trivial accident or incident at the most unexpected time when vagabonds, highwaymen, pimps, and pickpockets abounded and short-tempered city gentlemen wore swords, drank copiously, and cavorted about unlit streets, a prey to robbers and their own belligerent tempers. The antiquarian John Aubrey reports that while in London he was on more than one occasion nearly killed by irate gentlemen he had never met.[20]

Thomas Hobbes's view of life as "nasty, brutish, and short" clearly captured the experiences of a great many people living in early modern Europe and America. For in addition to disease and man's inhumanity to man, nature itself was frequently a cause for alarm. Any turn in the weather could spell scarcity and starvation because roads were few and bad and transportation precarious. The sixteenth and seventeenth centuries were marked by what has been described as a little ice age, which significantly reduced food production, increasing the number of famines.[21] The reason towns had walls was to keep out the human and animal dangers lurking beyond their perimeters. The dread of fire was a constant source of anxiety in an age when wooden bridges, wooden ships, wooden buildings, and even wooden chimneys were the norm. Without police, fire brigades, or insurance companies there was no clear way to prevent or mitigate any of these dangers.[22]

Yet for all the poverty, insecurity, and superstition, the early modern period produced a stunning galaxy of writers, artists, philosophers, and scientists who laid the foundation for modern culture and modern science. To paraphrase Alfred North Whitehead, we are still living off the capital accumulated by early modern intellectuals. This being the case, the question arises as to what prepared the way for the emergence of such talent, innovation, and invention during a period that was in major respects so dismal in both attitudes toward life and the actual living of it? Answering this question is a major goal of this book, but what should be stressed at the outset is that the early modern period, and particularly the seventeenth century, was a great age of transformation, to borrow Polyani's phrase.[23] The period marked the emergence of more modern economic and political systems as proto-capitalist forms of enterprise developed and nation states with increasingly centralized governments emerged. In addition, one of the great migrations occurred as Europeans colonized the New World and were brought face to face with diverse peoples with different histories, customs, and religions. All this contributed to a crucially important transformation, namely the way Europeans viewed themselves and the world. Paul Ricoeur claims that the Fall is "the central myth of western culture,"[24] a statement that applies particularly well to the early modern period, when discussions about the Genesis story of the Fall and the effects it had had on human nature were at their height. The sixteenth and seventeenth centuries have justifiably been called the "Age of Augustine" and "Augustinian anthropology" because of the harsh and unflattering view of human nature prevailing among both Protestants and Catholics.[25] By the end of the seventeenth century, however, a radical transformation had occurred as a new and positive view of mankind emerged that flatly contradicted Augustine, arguing instead that human beings were basically good and innocent until corrupted by society or other humans.

Augustine had originated the term "original sin" and claimed that as a result of the Fall human nature was "wounded, hurt, damaged, destroyed."[26] This was the view accepted by Lutherans, Calvinists, and many Catholics in the early decades of the Reformation. Not only had the Fall made it impossible for humans to act morally, but it had irreparably damaged Adam's intelligence and ability to reason. Because original sin was passed on to all of Adam's descendants through the act of procreation, every individual was consequently condemned to ignorance. But how extensive was this ignorance? Was it truly irremediable or was there some way to restore mankind to the prelapsarian state originally enjoyed by Adam? Peter Harrison argues that these questions were foundational to the pursuit of science in the early modern period: every theologian and natural philosopher grappled with the impact of the Fall on the human mind. As Harrison demonstrates, the different strategies and scientific schemes put forward in the early modern period for improving knowledge largely depended on their authors' assessment of the consequences of the Fall.[27] The opening lines of Francis Bacon's *Great Instauration* underscore this point with its initial question: "Whether that commerce between the mind of man and the nature of things . . . might be restored to its perfect and original condition, or if that may not be, yet reduced to a better condition than that in which it is now."[28] The answer Bacon gives, along with many of his contemporaries and successors, was a resounding "yes."

According to the conventional wisdom of the age, Adam had been created with an unblemished intellect and God had revealed to him complete knowledge of the natural world, as his naming the animals revealed. For to name things correctly Adam had to comprehend their essential natures. In paradise words were therefore synonymous with things, and because of this isomorphism Adam and Eve communicated with absolute clarity and without the slightest possibility of misunderstanding. It was this lost linguistic Eden that many people were anxious to recover and, along with it, the mental acuity of Adam. This explains why so many philosophers, theologians, and natural philosophers (as scientists were called at the time) devoted so much time to subjects we would not consider strictly scientific, such as epistemology and methodology as well as attempts to recover or recreate the perfect language spoken in Eden. In the view of many natural philosophers a perfect language was key to regaining the dominion over nature lost by Adam. As a result of these speculations and attempts to restore man to Adam's original perfection, by the end of the seventeenth century what might be described as an "anthropological revolution" had occurred: a more optimistic view of human nature emerged and along with it a positive attitude toward life and the ability of humans to change and improve their world and themselves. The following chapters will attempt to explain the reasons for this profound revolution in mentality and outline the key roles that religion, magic, and science played in this process.

The primary goal of this book is therefore to allow readers a glimpse into the lives of early modern Europeans and Americans to understand how and why they developed a more confident attitude about their ability to cope with events from the most trivial to the most traumatic. But a further goal is to focus attention on many of the issues debated by historians of the early modern period. Up to the 1960s, for example, the prevailing wisdom among scholars was that modern science emerged when, and only when, rationalist enlightenment philosophers rejected religion, magic, and esoteric or occult thought of any kind. In his *Introduction to the History of Science*, which was still required reading when I went to college in the 1960s, George Sarton made this claim abundantly clear when he declared at the very beginning of this three-volume work that the study of magic played no part in the history of science:

> The historian of science can not devote much attention to the study of superstition and magic, that is, of unreason, because this does not help him very much to understand human progress. Magic is essentially unprogressive and conservative; science is essentially progressive; the former goes backward; the latter forward.[29]

This view was radically challenged with the publication of Frances Yates's *Giordano Bruno and the Hermetic Tradition* (1964) and a succession of scholarly books arguing that religion and magic were not antithetical to science but in many cases actually promoted it. According to these historians, it was out of the confusing mix of mysticism, magic, religion, and science that a positive view of human beings developed, along with the conviction that science is a noble tool in the inevitable march of progress. But while the idea of progress and an optimistic view of the power and perspicacity of human beings characterized some aspects of what is conventionally described as enlightenment thought, this is not the whole story. As John Hope Mason remarked in 1982, the eighteenth century is no longer what it supposedly once was, a period of reason and order sandwiched between religious excess, on the one hand, and industrial and urban turmoil on the other. It now appears more conflicted, more somber, and more religious than once imagined. The idea made famous by Peter Gay that the Enlightenment represented the rise of modern paganism, is no longer tenable.[30] Gay could only make this claim because he located the Enlightenment in France, where the attack on religion was undeniable. But as scholars have convincingly shown, the Enlightenment took many different forms, reflecting the vastly different political, religious, and social environments in which it occurred, and in these various enlightenments religion played a crucial role, as indeed it did even in France. Thus, instead of seeing the Enlightenment as the first stage in the inevitable march to a secular modernity, Adorno and Horkheimer's famous thesis in *Dialectics of Enlightenment*, scholars now view it as a period in which new forms of media combined with new institutions to create a cacophony of competing and conflicting information supporting a proliferation of religious and political ideologies from the most conservative to the most radical.[31] Gay was therefore correct in illustrating the way skepticism, anti-clericalism, and irreligion proliferated during the Enlightenment, but he failed to appreciate how the battle between Catholics and Protestants also encouraged the revitalization of religious institutions, providing the laity with new forms of piety and new avenues for religious expression, all of which affected scientific thought and practice.

Just as our view of the Enlightenment has undergone considerable revision during the past half century as historians dug deeper into sources, exposing alternate currents of esoteric, magical, and religious thought, so too has our understanding of the early modern period. Scholars have rejected the once reigning view of Max Weber that the Reformation marked a decisive stage on the road to modernization and secularization as Protestantism encouraged the "disenchantment" of the world and modern forms of rational and scientific thought. What now seems certain is that the process of secularization took longer and was more complicated than Weber thought and that Catholicism contributed to the process as well. Furthermore, recent scholarship has brought enchantment back into the modern world in very suggestive ways.[32] A similar reevaluation has led scholars to question the once popular view of Alexandre Koyré that the Scientific Revolution occurred during the few decades before and after 1600, when the closed world of ancient Greeks and medieval Europeans was supposedly replaced by an infinite universe.[33] Along the same lines, the idea that the Scientific Revolution consisted of the acceptance of the so-called "Mechanical Philosophy," in which all that existed in the material world were atoms or corpuscles in constant motion, has given way to a new view in which atomism, Aristotelianism, and vitalism, together with a panoply of religious, occult, and esoteric theories, are all recognized as contributing to both the emergence of modern science and reactions against it. It was this recognition that led John Maynard

Keynes to describe Newton as "the last of the magicians" and "not the first of the age of reason" on the grounds that Newton spent far more time reading alchemical texts and interpreting biblical prophecies than he ever did on the astonishing discoveries in physics for which he is justly famous.[34] More recent scholarship confirms Keynes's insight by demonstrating the impact Newton's alchemical studies and heterodox religious views had on his physics.

An even more drastic assault on established scholarship occurred when the question arose about whether it was even possible to speak of a Scientific Revolution.[35] Steven Shapin begins his book, *The Scientific Revolution*, with the following ironic caveat, "There was no such thing as the Scientific Revolution, and this is a book about it." These radical reevaluations in the historiography of science were largely the result of the twentieth-century shift in interest from "Great Men" to the common man, woman, and child and to a new interest in popular culture as a whole. They were also due to a growing fascination with subjects like magic, the occult, alchemy, and millenarianism, a fascination reinforced by the atrocities of World War I and II, Nazism, Communism, and the Holocaust. Such stark examples of seeming irrationality made scholars more attuned to similar instances in the more distant past and to the role these played in shaping both our enlightened and unenlightened history. As a result a far more nuanced picture of early modern life has emerged, in which the boundaries between religion, magic, and science are in many cases obliterated or blurred beyond recognition and clear distinctions between Catholic and Protestant views of magic and science have proved untenable.

Whether Protestant or Catholic, early modern people lived in a magical universe infused with the supernatural. For the overwhelming majority of people, and this included the wealthy and the educated, witches, demons, angels, fairies, and ghosts were "real" beings, who interacted with people in their everyday lives. But while traditional ways of thought continued to maintain their hold, new ideas were emerging that questioned and in many cases rejected conventional assumptions. As we listen to news reports on the radio or television, we hardly think about what the word "news" really means. The term first came into common usage in the sixteenth century and referred to the astonishing amount of new information flooding Europe.[36] While for some people the barrage of data about unheard of plants, animals, and peoples was stimulating and exciting, others found it disconcerting. John Donne was not alone as he lamented a fragmented world in his poem "The Anatomy of the World": "Tis all in peeces, all coherence gone." The Ptolemaic-Aristotelian worldview with its tightly knit correspondences and interlocking categories could not accommodate the profusion of new and strange identities. As William Ashworth notes, "Anteaters and sloths do not appear in Erasmus or Alciati or Piero Valeriano; they are missing from all the writings of antiquity. They come into the Old World naked, without emblematic significance."[37] In 1623 the Swiss physician and botanist Gaspard Bauhin listed six thousand terminal taxa; in 1676 the English naturalist John Ray expanded the list to eighteen thousand, and no one stopped counting after that.[38] The old categories collapsed under the weight of so much new and indigestible information, leaving people adrift on unchartered intellectual seas. Richard Popkin spent a great part of his academic life describing the "skeptical crisis" that characterized this period of transformation as some individuals came to the conclusion that there are no absolute truths ordained by a supernatural, providential God, and consequently that man, in essence, makes himself, his values, and his society.[39] As one might expect, to counter such skepticism an unprecedented concern with order and orthodoxy arose among those dedicated to refuting such socially unsettling, even dangerous, suppositions. But by the end of the fifteenth and increasingly during the next three centuries a growing number of people began to harbor the idea taken for granted by Betrand Russell in the twentieth—that human existence has no extrinsic rhyme or reason, no supernatural purpose, plan, or goal:

> That man is the product of causes which had no prevision of the end they were achieving; that his origin, his growth, his hopes and fears, his loves and his beliefs are but the outcome of accidental collocations of atoms; that no fire, no heroism, no intensity of thought and feeling can preserve an individual life behind the grave; that all the labours of all the ages, all the devotion, all the inspiration, all the noonday brightness of human genius are destined to extinction in the vast death of the solar system, and that the whole temple of man's achievement must inevitably be buried beneath the debris of a universe in ruins—all these things, if not quite beyond dispute, are yet so nearly certain that no philosophy which rejects them can hope to stand.[40]

Russell's confident assertiveness belies the fact that most people could not, and still cannot, accept this bleak assessment of human life. As will become apparent in the course of this book, the concept of a meaningful universe died hard, or, for most people, never died at all.

If William Bouwsma is correct in defining culture as "a mechanism for the management of existential anxiety"—and I believe he is—at no time was such management more necessary than in the sixteenth, seventeenth, and eighteenth centuries. The early modern period was one of great anxiety fueled by the major social disruptions occurring as feudalism gave way to capitalism, agrarian life to increasing urbanization and industrialization, and regionalism to nationalism. While these transformations had profound effects on the lives of ordinary people and provided multiple causes for anxiety, they were not all negative. In fact, as previously mentioned, one of the major threads weaving through the following narrative is the increasingly optimistic assessment of human life that develops over the course of the period under investigation and the positive role that scientific developments had in fostering such optimism.[41] Alexander Pope's famous couplet "Nature and Nature's laws lay hid in night:/ God said, Let Newton be! and all was light" encapsulates this more positive view and the effect that Newton's scientific discoveries had in promoting a heroic view of science and scientists.

Newton's exalted reputation raises further important issues about the early modern period that need to be addressed. One of the most pressing is to understand how it was possible that the worst witch hunts in the long history of European violence, persecution, and warfare occurred during the lifetimes of such stellar scientists as Kepler, Galileo, Descartes, and Newton, and not during the Middle Ages as so many people mistakenly believe.[42] And why were women targeted as witches in far greater numbers than ever before at this particular time? Although at first glance witch hunting might seem antithetical to the kind of intellectual developments connected with the Scientific Revolution, it is more accurate to understand the witch hunts as an aspect of the transition from the pre-modern to the modern world, as an exemplary case, in fact, of a society undergoing religious, economic, and social change and the special way this kind of change affects women. Witches were certainly targets of hatred and persecution in earlier centuries, but their persecution intensified between 1550–1680 because they came to epitomize the disorder of the period. Their punishment, indeed, extermination, was considered by many religious and political authorities an essential element in the restoration of order and stability.

Witch hunts went hand in hand with other attempts to control the thought and behavior of individuals living in increasingly diverse and religiously divided communities. They can and should be seen as indicative of the major cultural transformation that prepared the way for the emergence of a modern, urban, and industrialized society. Witch hunting also played a crucial role in the evolving attitudes toward magic and in determining the boundaries between the natural and the supernatural. The investigation of phenomena such as witchcraft and the attempts to substantiate the existence of fairies and elves, the efficacy of magical spells, the reality of miracles, and the effectiveness of exorcism, along with a host of similar issues on the borderline between religion, magic, and science, were therefore an integral part of the Scientific Revolution. Believing in witchcraft because it was mentioned in the Bible and confirmed by ancient sources may have been accepted as adequate proof by some people, but for increasing numbers second-hand testimony, even from the Bible, was no longer sufficient. Direct observation was the only path to truth. Starting with Bacon, natural philosophers insisted that, unlike their predecessors, they did not frame hypothesis or spin theoretical webs; they simply presented the "facts" and allowed them to speak for themselves.[43] It may be somewhat surprising to realize that this emphasis on "facts" united natural philosophers with demonologists and witch hunters, a state of affairs that once again underlines the inseparable nature of religion, magic, and science at the time. In major respects these groups were not at loggerheads; they were literally on the same page, so to speak, sharing the same concern with overcoming skepticism, relativism, and the atheism they were convinced stemmed from both. The attempt made by philosophers and scientists as well as theologians and witch hunters to distinguish fact from fiction had profound effects on the way people thought, especially in regard to causality. Statistical analysis and the idea of probability come into their own during the early modern period; both are characteristic of modern rather than pre-modern thought, and both signaled the decline of a magical world view among the educated.

A final question that will be addressed concerns the role and extent of atheism and heresy in early modern Europe. Many theologians and intellectuals were convinced that religion was under attack from rationalist enlightenment philosophers and that this attack threatened to unravel the fabric of society. What is particularly interesting is that many of the most prominent natural philosophers, men at the center of the Scientific Revolution like Robert Boyle and Samuel Clarke, believed this. How true was it? It was actually both true and false, but the very idea that such an attack was occurring is indicative of the unsettling transformations that occurred in the aftermath of the Reformation and the destabilizing effects these have had ever since. As Michael Hunter has perceptively argued, the great fear of atheism that begins to loom large at the end of the seventeenth century expressed an even greater fear of the secularization that was occurring as the church lost authority over areas previously under its control, such as law courts, social welfare, and the regulation of church building.[44] "Down with the Enlightenment" became the rallying cry of conservatives and reactionaries from the end of the eighteenth century onward and continues in full force today. From Horkheimer and Adorno to the postmodernist critique of John Gray and contemporary conservative thinkers, the enlightenment rejection of religious authority and emphasis on reason and human autonomy have been held responsible for undermining ethics and morality to such an extent that colonialism, imperialism, totalitarianism, fascism, and even the ecological crisis inevitably followed. Interestingly enough, the attack on the Enlightenment comes from the left as well as the right, from postmodernists who view the claim for reason, tolerance, and universal brotherhood as a cynical ploy to mask Western domination and exploitation. Thus hostility toward the Enlightenment permeates both conservative and postmodernist thinking—strange bedfellows one might think—and in doing so obscures the important contribution enlightenment thinkers made to establishing the values of individual liberty, democracy, toleration, equality, and universal rights. It was not the Enlightenment, but the rejection of enlightenment ideals by certain romantics, nationalists, religious conservatives, reactionaries, and disappointed radicals that laid the foundations for the fundamentalism, nationalism, cynicism, and brutality that fueled the holocausts of the twentieth century as well as the religious terrorism of the present day.

From the vantage point of the twenty-first century the early modern period is both strange and familiar. Like the contemporary world it was a time of tremendous confusion, unparalleled questioning, and the rejection of traditional beliefs, all of which led to new and unprecedented discoveries in science, the arts, and the humanities. The period between 1450 and 1800 witnessed the creation of a world in which change, progress, an appreciation of science, faith in the value of education, commitment to tolerance, and respect for the individual came to the fore. At the same time it marked a hardening of attitudes on the part of those groups and individuals who rejected change in favor of the status quo, who distrusted science, feared education, and privileged their view of the common good over individual rights and freedom. The early modern conflict between the so-called "Ancients" and "Moderns" anticipated our own era's conflict between conservatives and liberals, and this conflict is no less contentious than it was in these earlier centuries. Such conflict is, perhaps, an indelible and ineradicable aspect of Western modernity itself.[45]

CHAPTER 1

All Coherence Gone

Guilio Camillo (1480–1544) was one of the most famous men in the sixteenth century. Known as "the divine Camillo," he constructed a memory theater that attracted the attention of the French King Francis I, who helped finance its construction. The fame of this theater spread throughout Europe as it made its way from Paris to Venice, attracting crowds of visitors. Modeled on Solomon's Temple, the theater presented itself as a repository of all knowledge, ancient and modern. Stretching outward from the stage were seven gangways, each dedicated to one of the seven planets. Arranged along each gangway were the arts, sciences, elements, entities, and activities associated with that planet from the moment of creation to the present. In a letter to the Dutch Humanist Desiderius Erasmus (1466–1536) Viglius Zuicheus describes the theater, evidence in itself of the interest it generated:

> The work is of wood, marked with many images, and full of little boxes; there are various orders and grades in it. He gives a place to each individual figure and ornament . . . He calls this theatre of his by many names, saying now that it is a built or constructed mind and soul. . . . He pretends that all things that the human mind can conceive and which we cannot see with the corporeal eye, after being collected together by diligent meditation may be expressed by certain corporeal signs in such a way that the beholder may at once perceive with his eyes everything that is otherwise hidden in the depths of the human mind.[1]

What excited so much attention was the vision Camillo's theater offered of the universe in miniature and the promise that by standing on the stage and contemplating the theater's contents viewers would absorb an encyclopedic knowledge of the world that could be used for both personal and public advantage. This encyclopedic knowledge of the world already existed in the human mind, but Camillo's theater made it clearer and more immediately accessible. Seeing and absorbing everything before them, those privileged to stand on Camillo's stage literally had the world in their hands, and with this knowledge they could manipulate it and its inhabitants. Although the theater was never completely finished (and indeed disappeared), it remains a monument to Camillo's confident assessment of the powers of the human mind to comprehend and control the world.

The fame of Camillo's theater was matched by the legends circulating about its architect. According to one, while in Paris Camillo had accompanied several distinguished gentleman to view some wild animals. A lion escaped and came toward the group, at which point the story is picked up by one of Camillo's companions:

> The gentlemen were much alarmed and fled hither and thither, except Messer Giulio Camillo who remained where he was, without moving. This he did, not in order to give proof of himself, but because of the weight of his body which made him slower in his movements than the others. The king of animals began to walk round him and to caress him, without otherwise molesting him, until it was chased back to its place. What will you say to this? Why was he not killed? It was thought by all that he remained safe and sound because he was under the planet of the sun.[2]

That Camillo could remain unharmed by a wild beast was, as this account suggests, a cause of wonder and taken as a sign that he enjoyed special planetary and divine protection.

Camillo's vision is not ours, and to understand both the rationale and fame of his theater requires a step back in time and an appreciation of the way people viewed the world before the effects of the Copernican Revolution were fully absorbed. We take the Copernican sun-centered universe for granted, but the geocentric world of the Aristotelian-Ptolemaic system that preceded it was in every respect more comprehensible and, one might even say, more "homey" and comfortable for human beings than what came afterward. For the pre-Copernican universe was finite and more

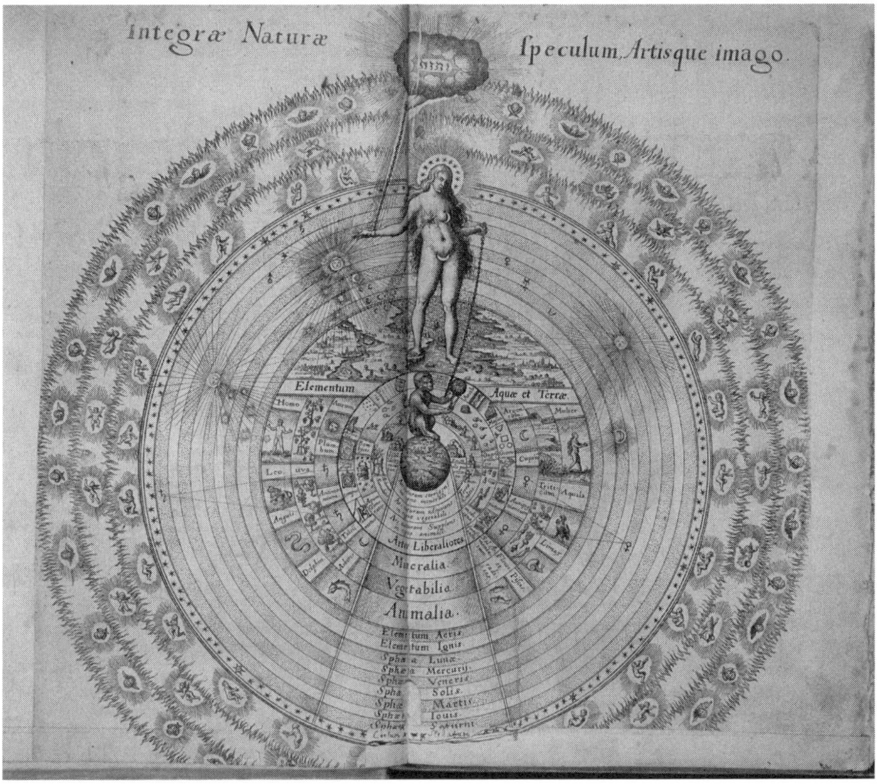

Microcosm-Macrocosm. (From Robert Fludd, *Utriusque cosmi maioris scilicet et minoris metaphysica, phyisca atque technica historia*. . . . Oppenheim, 1617–1618. Vol. 1, fol. 4–5. Wellcome Library, London.)

suited to human proportions than the infinite world of post-Copernican cosmology. Everything in the pre-Copernican world had its ordained place while simultaneously participating in a web of sympathies that linked heaven to earth and each thing to every other. Through a series of correspondences, stars and planets were connected to men, animals, plants, minerals, and metals, and every existing thing had a distinct essence and meaning that could be "read" by discerning minds. Louis XIV was suitably described as the "Sun King" and his image associated with gold and lions because the sun was the preeminent planet, just as the lion was the king of beasts and gold the most precious of metals. Individual entities were part of an all-encompassing, organic whole or "Great Chain of Being," as it was

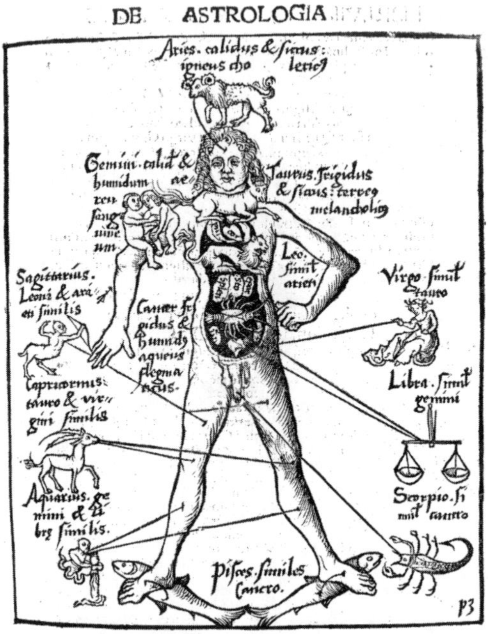

Zodiac Man. (From Gregor Reische, *Margarita philosophica. . . .* Friburgi, 1503. Wellcome Library, London.)

often called, in which each link contained traces of the previous one. As the microcosm that reflected the greater world, or macrocosm, human beings were essentially the world in miniature. "Zodiac Man" provides a good illustration of this kind of analogous thinking, for in this image each part of the human body is connected to the various signs of the zodiac. The planets and stars not only influenced man's actions and destiny but determined his physical health as well. These physical connections between the heavenly and human worlds explain the theory of the human temperaments, which was a basic aspect of the medical theories of the second-century Roman physician Galen, whose work was still influential in the early modern period. A jovial, optimistic, or sanguine temperament reflected the beneficial influence of the sun and solar forces—the very forces that saved Camillo from the lion—while a melancholy temperament indicated the baneful influence of Saturn, and an excess of choler, or anger, was due to Mars.[3]

Scholars have described this way of envisioning the universe as an "emblematic" worldview, according to which the universe was a vast open book with every entity linked to every other through a network of signs and symbols that, while hidden, were accessible to the human mind. An example from the work of the great Swiss naturalist Conrad Gesner (1516–1565) illustrates this mode of thinking: to fully understand the nature of a peacock, an anatomical description of the bird was insufficient. One had to draw on history, myth, and etymology and take into account all the associations, affinities, and sympathies linking the peacock to the rest of creation. Combing through fables, emblems, adages, and proverbs was therefore just as much a part of natural history as observing nature itself. William Ashworth, who has written extensively on the emblematic worldview, explains: "To know the peacock, as Gesner wanted to know it, one must know not only what the peacock looks like but what its name means, in every language; what it symbolizes to both pagans and Christians; what other animals it has sympathies or affinities with; and any other possible connection it might have with stars, plants, minerals, numbers, coins, or whatever." The idea that anyone could understand the nature of a peacock simply by studying its physiology was, as Ashworth emphasizes, "a notion completely foreign to Renaissance thought."[4] According to the emblematic view of the world, each entity possessed a "signature" that enabled human beings to comprehend its nature, significance, and use. This was a guiding principle in the homeopathic medical theories of Martin Luther's contemporary, the revolutionary physician Paracelsus (1493–1546) and his followers: anconite is a good medicine for the eyes because its seeds look like eyes; walnut meat cures internal head ailments because it looks like the brain.[5] In other words, the signature of an entity enabled the naturalist to "see" what that entity resembled and therefore to know what it was like and how it could be used.

Resemblance played a constructive role in organizing information in the pre-Copernican universe as well as in guiding the exegesis and interpretation of texts. Things were organized into categories according to their similarity to other things. Resemblance and analogy enabled people to move from knowledge of visible to invisible things.[6] Furthermore, as we have seen in the case of Camillo's theater, an emblematic worldview was predicated on the assumption that everything in the human mind, which itself replicated everything in the world at large, could be expressed by certain corporeal signs. Thus, God had so constituted the human mind that it contained a mirror image of every created thing. This assumption led to the optimistic conclusion that human beings could obtain absolute knowledge of the world. This conviction was a basic aspect of Paracelsian thought:

> But we men discover all that is hidden in the mountain by signs and outward correspondences; and it is thus that we find out all the properties of herbs and all that is in stones. There is nothing in the depths of the seas, nothing in the heights of the firmament that man is not capable of discovering. There is no mountain so vast that it can hide from the gaze of man what is within it; it is revealed to him by corresponding signs.[7]

Paracelsus's statement mirrored the optimistic assessment of human nature characteristic of his predecessor, the Italian philosopher and Humanist Giovanni Pico Della Mirandola (1463–94), the *wunderkind* of the Florentine Academy established by Cosimo di Medici. Drawing on the ancient texts supposedly written by an Egyptian contemporary of Moses, Hermes Trismegistus (Thrice-Great Hermes), texts that had been recently rediscovered and translated at the Medici court, Pico claimed that he finally understood why man was the "most fortunate" of all the creatures and worthy of admiration. Man was a "miracle" because his creator had endowed him with the capacity to become whatever he wished. He could scale the heights and plumb the depths, becoming an angel or beast or even divine:

> We have given you, O Adam, no visage proper to yourself, nor endowment properly your own, in order that whatever place, whatever form, whatever gifts you may, with premeditation, select, these same you may have and possess through your own judgment and decision. The nature of all other creatures is defined and restricted within laws which We have laid down; you, by contrast, impeded by no such restrictions, may, by your own free will, to whose custody We have assigned you, trace for yourself the lineaments of your own nature. I have placed you at the very center of the world, so that from that vantage point you may with greater ease glance round about you on all that the world contains. We have made you a creature neither of heaven nor of earth, neither mortal nor immortal, in order that you may, as the free and proud shaper of your own being, fashion yourself in the form you may prefer. It will be in your power to descend to

> the lower, brutish forms of life; you will be able, through your own decision, to rise again to the superior orders whose life is divine.[8]

It is sobering to realize that these words were written thirty-one years before Luther nailed his ninety-five theses to the cathedral door in Wittenberg, thus ushering in the Reformation, one of the bloodiest, most intolerant, and pessimistic periods in European history. But it is even more sobering to realize that shortly after he had written his "Oration" and before his premature death in 1494, Pico became a follower of the fiery Dominican preacher and moralist Girolamo Savonarola (1452–98) and planned to enter a monastery. The condemnation of Pico's ecumenical views by the Pope and his brief imprisonment at the instigation of Catholic officials dashed Pico's youthful enthusiasm for an inclusive, tolerant religion based on the goodness and perspicacity of man, and he turned instead to the ascetic, apocalyptic teachings of the Dominican friar.

The optimism characteristic of Pico's "Oration" would be eclipsed for the next century and a half, never disappearing entirely, but taking a back seat as a darker view of human nature and human life emerged in the writings of Protestant reformers, who found their inspiration in Augustine (354–430). Augustine lived during the decline of the Roman Empire, when social and political structures were collapsing and the ideologies and social conventions that had previously made sense no longer appeared to apply. Augustine's emphasis on the baseness and corruption of human nature reflected the chaotic times in which he lived.[9] His views resonated with the two greatest leaders of the Protestant Reformation, Martin Luther and John Calvin, who themselves lived in a time of religious chaos and confusion. The pessimism of the sixteenth century has been amply documented. Enormous cultural shifts occurred in Europe from the mid-fourteenth century onward as a result of increasing contact with non-Europeans and the effects of printing and scholarship in raising questions about medieval theology and philosophy. When combined with the profound social and economic changes accompanying the emergence of an increasingly urban, proto-capitalist culture and the wars of religion precipitated by the Reformation, one finds a perfect storm of factors at hand to shatter the teleologically and hierarchically ordered Aristotelian-Ptolemaic cosmos.[10] With the demise of this worldview went the framework that had allowed Europeans to understand the world, their place in it, and their purpose and identity for centuries. Although many people remained blissfully ignorant of these developments, it is not an exaggeration to say they had tragic effects. However disastrous the Fall and the expulsion from the Garden of Eden was in the minds of believing Christians, the fall from the pre-Copernican into the post-Copernican universe was even more traumatic; for it was not simply a matter of the expulsion of sinful humans with diminished capabilities into a less-than-perfect world, but the emergence of novel and unsettling questions about the nature and reality of this world and the humans who inhabit it. Fast on the heels of these existential conundrums came the horrifying possibility that not only was knowledge in any absolute form unattainable but the very existence of a stable, substantial, and reliable knower had become problematic.

Jonathan Z. Smith has emphasized the importance of place in a culture's and an individual's self-perception. Because social change is inseparable from symbolic change, the question people ask and need to answer to ensure a stable existence is "where do I stand?" As Smith says: "At the heart of the issue of change are the symbolic-social questions: What is the place on which I stand? What are my horizons? What are my limits?"[11] These were precisely the questions that preoccupied and, in many cases, terrified individuals in the early modern world as they cleared away the debris of outmoded symbolic structures and struggled to build new ones to accommodate a changing world. To quote Smith once again: "To change stance is to totally alter one's symbols and to inhabit a different world."[12] This was the difficult "position" Luther found himself in at The Diet of Worms (1521), when his "stand" against the Catholic Church promised a fundamental alteration in the relationship between church and state, priests and parishioners, and individuals and God. As Luther said: "I am bound by the Scriptures I have quoted and my conscience is captive to the Word of God. . . . Here I stand. I can do no other. May God help me. Amen."[13] What caused individuals like Luther and so many others to change their stands on so many key issues was the increasingly obvious lack of coherence in the symbolic systems governing contemporary culture, a disjunction revealing the lack of "fit" between the various elements within the prevailing systems of classification, be they social, intellectual, economic, political, or religious.

What thinking Europeans were forced to confront in the early modern period was radical change, and this confrontation required altering the way they viewed the universe, their culture, and their fellow men and women. No longer living in an earth-centered universe in which heaven was suitably situated in the perfect, unchanging, and ethereal heavens and hell in the bowels of the earth, the darkest, densest, and, by implication, foulest part of the physical universe, post-Copernican men and women found themselves inhabiting one planet among many, hurtling through space at incredible speeds. Not only was the formerly stable earth spinning around its own axis as it simultaneously traveled around the sun, but all this took place in what now appeared to be an infinite universe of unfathomable vastness. The idea of space changed radically in this new universe, where there was no longer a clear sense of "up" or "down" and hence no commonsensical place for heaven or hell. Furthermore, if earth had no privileged position but was merely one of many planets, did that mean that Christ's incarnation, crucifixion, and resurrection had to be repeated on every planet in every solar system in an infinite universe? Pondering such questions led the Englishman John Edwards to conclude that if Copernicus were correct, "this terrestrial Globe is a despicable spot, a speck, a Point in Comparison of the vast and spacious conjeries of the sun and fixed Lights."[14] The existential dread such vastness stirred up in many minds comes out clearly in Pascal's famous comparison of man to a fragile reed in a vast, uncaring universe.[15]

The contrast between the view of man presented by Pascal and Pico in his "Oration" is striking. While for Pico man is a marvel, a chameleon able to shape himself freely according to his own will, for Pascal he is a monster and a chimera, a grotesque, chaotic assemblage of incompatible parts, wishes, and desires: "What a chimera, then, is man! What a novelty! What a monster, what a chaos, what a subject of contradiction, what a prodigy!" There is no hint of wonder in Pascal's prose when it comes to his assessment of the human condition as there is in Pico's. Pascal's tone mixes irony and pity as he juxtaposes man's grandiose pretensions against his meager accomplishments. "Judge of all things, imbecile worm of the earth; depository of the truth, a sink of uncertainty and error, the pride and refuse of the universe!"[16] One might ask how a single being can be both a "depository of the truth" and a "sink of uncertainty and error" as well as "the pride of the universe" and "its refuse." The answer is that human beings lack consistency and wholeness; they are monstrous mixtures of warring fragments. Pascal cuts man down to size: for all his pretensions to knowledge, pontifications about morality, and aspirations to grandeur, he is "nothing" in comparison to nature: "What is man in the midst of nature? A nothing in comparison with the infinite, an all in comparison with nothingness: a mean between nothing and all."[17] The single positive thing about men is that they think: "Man is but a reed, the weakest in nature, but he is a thinking reed."[18] How positive thinking really is, however, is debatable because a thoughtful person would presumably reach the same dismal conclusions as Pascal. In addition, thinking isn't capable of doing humans much good inasmuch as "[a] breath of air, a drop of water, suffices to kill man."[19]

Not everyone was as horrified as Pascal by the notion of infinity and the vastness of the universe. Pascal's existential dread was typical of those who lamented the loss of the "closed" and "bounded" world of the Aristotelian-Ptolemiac universe, in which one's place was clearly established and "walls," both actual and metaphorical, kept out whatever appeared to be beyond the social and intellectual order. A coterie of adventurous spirits existed however, who, like Giordano Bruno (1548–1600), the Italian Dominican friar burned at the stake for heresy, relished the idea of a vast and "open" universe and longed to escape the restrictions imposed by social and religious conventions. By the end of the seventeenth century these voices became more numerous, and there was a reemergence of the kind of optimism characteristic of Pico. But this is to get ahead of the story. Such optimism only rose from the ashes of the discredited Aristotelian-Ptolemaic worldview, a demise that needs to be described in greater detail.

In an important article William Bouwsma argues that the kind of pessimistic view of human nature characteristic of Pascal was a defining aspect of the growing anxiety that developed during the late medieval and early modern periods. The Black Death, the depressed economy, the transition from a corporate to an individualistic society, the contraction of Christianity in the East, and the disarray within the Catholic Church were all important factors in provoking this anxiety; but Bouwsma locates an even more fundamental cause in the transition from agrarian to urban society. Even today cities inspire revulsion on the part of those who see them as sources of sin, corruption, and social blight. Jeremiads against cities were common in late medieval and early modern Europe. Cities were credited with fostering greed and rapaciousness and rupturing bonds between family and friends, all in the name of profit. Urban life encouraged social mobility, breaking down boundaries and distinctions between individuals.[20] Many intellectuals went even further in their condemnation of contemporary life, consumed by the idea that the entire natural world was in a state of irreparable decay. The existence of irregular mountain ranges was taken, for example, as a sign of the earth's degeneration from its originally perfect spherical shape.[21] In his *Theory of the Earth* Thomas Burnet claimed that when formed the earth was smooth like an egg with no rivers, seas, or mountains: "it had the beauty of Youth and blooming nature, fresh and fruitfull, and not a Wrinkle, scar or fracture in all its body; no Rocks nor Mountains, no hollow caves nor gaping channels." Over time the sun caused the earth's crust to crack and waters were released so that now we "walk upon its Ruins."[22]

The perceived disorder in the natural and social worlds was exacerbated by the bitter religious conflicts of the period and the proliferation of new and, in many cases, startling information coming from both the New World and the Old. The flood of new ideas that arrived with both the rediscovery of classical texts and the discovery of new lands, continents, and peoples undermined traditional philosophical and scientific frameworks. Gabriel Harvey describes the instability created by the constant barrage of "news":

> All inquisitive after Newes, new Books, newe Fashions, newe Lawes, newe Offices, and some after newe Elements, some after newe Heavens, and Helles to. . . . [A]s of olde Bookes, so of aunciente Vertue, Honestie, Fidelitie, Equitie, newe Abridgementes: every day, freshe span newe Opinions: Heresie, in manners, grounded much upon hearsay: Doctors contemned: The Text knowen of Mosis, understood of fewe; magnified of all: practiced of none: the Divell not so hated, as the Pope: many Invectives, small amendment.[23]

The fascination with prodigies, apparitions, comets, monsters, amazons, hermaphrodites, and witches—in short, with everything "unnatural" and "abnormal"—was indicative of the profound anxiety caused by the destruction of existing categories under the weight of all this new information. The grand schemes of encyclopedic knowledge characterizing the Middle Ages based on correspondences binding every aspect of the physical world into an organic, hierarchical unity collapsed under the weight of what has been described by many scholars as an information explosion.

It is difficult for modern readers to realize that before the Renaissance and even later the idea that something was new or novel was not positive, but in many cases disconcerting, even threatening. Stephen Greenblatt has captured the profound psychological effect the New World had on Old World psyches. In the face of so much novelty, the ancient maxim *Nil admirari* no longer made sense:

> in the presence of the New World the classical model of mature, balanced detachment seemed at once inappropriate and impossible. Columbus's voyage initiated a century of intense wonder. . . . European culture experienced something like the "startle reflex" one can observe in infants: eyes widened, arms outstretched, breathing stilled, the whole body momentarily convoluted. But what does it mean to experience wonder? What are its origins, its uses, and its limits? Is it closer to pleasure or pain, longing or horror? . . . The expression of wonder stands for all that cannot be understood, that can scarcely be believed. It calls attention to the problem of credibility and at the same time insists upon the undeniability, the exigency of experience.[24]

Greenblatt illustrates this point by citing the French Huguenot pastor and member of the abortive French colony established at Rio de Janeiro in the 1550s, Jean de Léry (1536–1613), whose *History of a Voyage to the Land of Brazil* was published in 1578. Léry wonders how his readers can be made to "believe what can only be seen two thousand leagues from where they live; things never known (much less written about) by the Ancients; things so marvelous that experience itself can scarcely engrave them on the understanding even of those who have in fact seen them?"[25] The answer was that to believe in the reality of such marvels entailed questioning and eventually rejecting traditional ideas, categories, and philosophical and scientific schemes. Margaret Slaughter illustrates the "taxonomic confusion" caused by such a quantitative leap in available information in the particular case of botany. She cites the table of contents from Dodoens's herbal published in 1554. This makes absolutely no sense from a modern point of view:

> Book i . . . lists herbaceious plants in alphabetical order. Book ii is on flowers used in garlands and bouquets, or for their aroma, and also treats of the umbellifers. Book iii is on roots, vines, poisonous plants, and cryptograms. Book iv concerns cereals, leguminous plants, and those of the marsh and seashore. Book v treats of edible plants, and Book vi is about shrubs, trees, forest trees, and evergreens.[26]

As this attempt at classifying plants reveals, the old categories could not accommodate such unexpected and therefore incomprehensible information; consequently, things we now know to be completely unrelated are lumped together chaotically.

The explosion of information and subsequent breakdown of conventional categories were important in fostering a skeptical attitude to received ideas as well as an awareness of the way culture conditions beliefs. While increasing travel, trade, and conquest were instrumental in unsettling the mental horizons of Europeans, so too was the scholarly work of the Humanists. The discovery and translation of hitherto unknown Greek, Latin, Arabic, and Hebrew philosophical, scientific, and literary texts presented Europeans with a smorgasbord of new ideas and an array of philosophies that provided further alternatives to the basic Aristotelianism of the Catholic Church. The skepticism, relativism, and downright heresy stimulated by these newly discovered texts has been amply described by Frances Yates, D. P. Walker, Richard Popkin, and others. But it is useful here to give a few indications of how deep this skepticism went and how radically it undermined existing beliefs. Montaigne, for example, accepted a degree of relativity in religious matters that most people reject to this day. As he said, "We are Christians for the same reason that we are Perigordians or Germans."[27] Montaigne had epigrams from the ancient skeptic Sextus Empiricus (c. 160–210 CE) carved on the ceiling beams in his library. His ultimate answer to the most unsettling question of all, "Que sais-je?" ("what do I know?") was "Moi-même" ("myself"). Montaigne's disciple, Pierre Charron, extended skepticism to call into question the basic concepts of "human nature" and "natural law," both of which had traditionally enabled people the luxury of thinking they knew who and what they were:

> We can no longer recognize anything of nature in ourselves. If we had to say how many laws of nature there are, and what they ordain, we would be at a loss. The sign of a natural law must be the universal respect in which it is held, for if there was anything that nature had truly commended us to do, we would undoubtedly obey it universally: not only would every nation respect it, but every individual. Instead there is nothing in the world that is not subject to contradiction and dispute, nothing that is not rejected, not just by one nation, but by many; equally, there is nothing that is strange and (in the

> opinion of many) unnatural that is not approved in many countries, and authorized by their customs.[28]

The seedbed for this kind of radical skepticism had been sown earlier by Renaissance Humanists, whose linguistic skills and careful textual scholarship contributed greatly to the idea that language was not a mirror of reality but a social construct and that consequently the meaning of a text could be determined only through a thorough understanding of the historical context in which it had been written. Paul Kristeller was the first to emphasize the centrality of rhetoric in Renaissance Humanism.[29] The implications of such an emphasis on rhetoric were profound. Bouwsma comments, "Man, as the rhetorician, not man as a species but man in a particular time and place, becomes the measure of all things."[30] The awareness of Lorenzo Valla (1408–57) and other Humanists that language conditions thought laid the foundations for the modern concepts of historical change and development.[31] This, in turn, fostered the notion that all knowledge is relative to time and place. Modern philology originated in fifteenth-century Italy with Valla and Poliziano (1454–94), and with philological studies came the realization that languages developed and changed over time. This kind of philological analysis led Valla to dismiss as a forgery the so-called "Donation of Constantine"[32] and Budé to conclude that the medieval Italian tradition of jurisprudence associated with Bartolus was a chimera based on defective texts and a misunderstanding of Roman history. Their careful textual scholarship undermined the idea that eternal truths existed beyond time and irrespective of cultural conditions and traditions.[33]

The same lesson about the protean nature of language and its inherent instability was also an integral part of humanistic education with its stress on rhetoric. The emphasis on rhetoric in the education of early modern males fostered an aggressive, polemical outlook antithetical to compromise and consensus.[34] The abandonment in many quarters of medieval logic with its penchant for syllogisms and its replacement by Ciceronian topics and Quintilian's analysis of the techniques of argumentation encouraged the idea that probability, not certainty, was the goal of argumentation: to win was more important than to discover the truth. The new rhetoric of argumentation was admirably suited to the social and political developments that led to increasingly centralized governments during the early modern period and to the emergence of a new class of government officials and courtiers, whose ethos of *sprezzatura* (the art of concealing art) and whose very existence depended on agile and constant self-fashioning.[35] The study and practice of rhetoric thus revealed the way in which words could be used to create reality: "The whole humanist focus on language as a sociohistorical product implied what Valla sporadically inferred from it: that semantics is epistemology, that language does not reveal or reflect reality but constitutes it."[36] Yet, as Waswo demonstrates, this basic insight was rejected, ignored, or suppressed. He provides numerous examples from Vives, Erasmus, and others to illustrate their extreme reluctance to accept the logical consequences of the linguistic ideas they themselves accepted. "[L]ike a moth to the flame" these authors constantly return to the idea that language constitutes meaning, but they do everything in their power to escape the implications of this insight.[37] For the most part, humanist scholars continued to take an ahistorical view of language as a whole, even if they realized its historicity in the short term. For obvious reasons, they could not envision the tremendous sweep of history that began to emerge in the eighteenth century with incipient ideas of evolution. Relativism thrived in this later climate, and in many respects Romantics picked up where Renaissance and early modern scholars left off. The result was what Waswo describes as an "abortive intellectual revolution," which he believes lies at the very heart of the Renaissance.[38]

This inability of intellectuals to fully appreciate the implications of their own scholarship is one factor that has led some historians to take a more cautious position on the destabilizing effects of the "information explosion." In their view the power of tradition and authority was more durable than many scholars realize. It is undoubtedly true that Europeans were reluctant to abandon inherited structures of knowledge. Like every other group, they tended to interpret new information in the light of existing categories. An example of this tendency is Columbus's swift reassessment of his initial view of the "Indians" as modern inhabitants of the Garden of Eden, which for Columbus and most Christians was an actual place, lost but discoverable. Having initially thought this because of their nakedness, generosity, and innocence, upon further reflection Columbus concluded that what he had really discovered was a decadent civilization of naked savages, whose generosity was a sign of their stupidity. It was the rare European who could really entertain the idea of "the noble savage" in anything more than a rhetorical sense. As in Columbus's case the superiority of "Us" versus "Them" reasserted itself when challenged in any substantial way.

It is also true, as revisionist historians claim, that during the first centuries of printing traditional texts and images of an earth-centered cosmos continued to appear, suggesting that the rupture between old and new ways of thinking were not as clear-cut as many scholars imagine. For all the continuities, however, the increasing availability of printed texts, even when completely traditional, undermined the very traditions presented in them. As Elizabeth Eisenstein, Anthony Grafton, and others have pointed out, it was not so much the presentation of new ideas that shattered existing schemes of knowledge as the recognition that substantial contradictions existed in texts that had been read and accepted for centuries.[39] These contradictions and discrepancies were not as easy to spot in a culture dependent on manuscripts, because manuscripts were more expensive than printed books and therefore harder to collect and compare. With printing, collecting and comparing texts became much more widespread, a practice aided by new technological inventions. In 1588 Agostino Ramelli, an Italian craftsman, described a device called the book wheel that allowed a reader to display particular pages of up to seventy books at one time to compare their contents. Desk chairs with wheels appeared about the same time, allowing scholars to scoot back and forth from desks to library shelves in search of more material. The consequences of the enhanced ability to compare texts can be seen in Erasmus's edition of the gospels laid out on the printed page in contiguous columns. This layout brought home to readers as never before both the similarities between the gospels and, more importantly, their differences. The compilation and publication of the great polyglot Bibles of the sixteenth and seventeenth centuries offer further examples of the role that antiquarian and humanist scholarship, together with printing, played in undermining traditional beliefs and assumptions. Not only did the polyglots raise all kinds of issues about the exact meaning of scripture and consequently about the way they had been both translated and interpreted through the ages by Church Fathers and theologians, but they also raised questions about the institutional organization and rationale of the Catholic Church itself.[40]

It comes as a surprise to many people that before the Reformation the Bible was seldom read by ordinary lay men and women. The Catholic Church discouraged individuals from reading scripture, realizing what

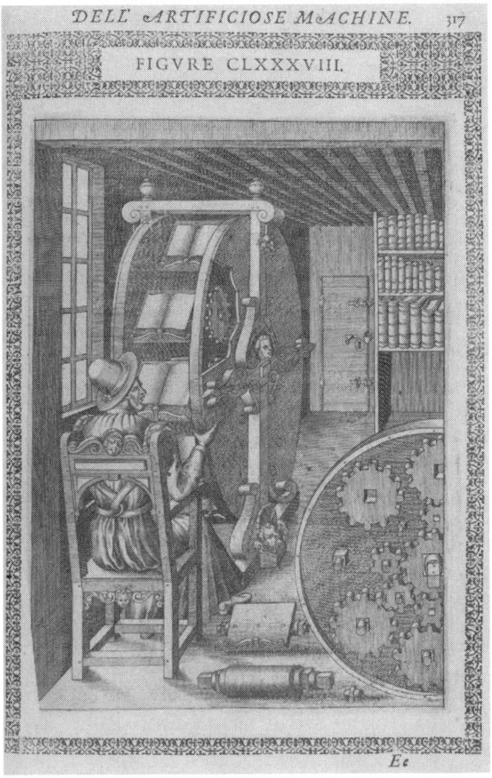

Agostino Ramelli's Book Wheel. (From *Le diverse et artificiose machine*. Paris, 1588. Wellcome Library, London.)

Luther and other Protestants would later discover to their chagrin, namely that without proper guidance the meaning of the scriptural text is neither clear nor self-explanatory. Scripture is filled with passages describing customs and rituals, not to mention objects and concepts, that are strange, jarring, even incomprehensible for readers coming to these texts long after they were written. What for example was an early modern reader to make of Lot's daughters' incestuous relationship with their father (Genesis 19) or with Jephthah's sacrifice of his daughter (Judges 11)? And what about entire texts like the *Song of Songs*, which most modern commentators now consider a secular love poem? Medieval theologians and exegetes dealt with such difficulties in the scriptural text by applying the prevailing fourfold interpretative scheme. If the literal interpretation of a text presented problems, three other interpretive options were available: the text could be read allegorically; it could be taken topologically to express a moral meaning; or it could be interpreted anagogically as a reference to what would happen at the end of time. In many cases, scriptural passages could be and were read in all four ways.

The fourfold interpretation of the text embodied the idea that scripture contained eternal truths valid for all times. While these truths might be expressed in an outdated or foreign idiom, they could be reconfigured through allegorical and symbolic interpretations in a way that became meaningful to contemporary readers. For example, for patristic and medieval exegetes the *Song of Songs* was not a secular love poem but a description either of the relationship between Christ and the soul or of Christ and the Church. The Garden of Eden was not simply a geographical place with certain topological features, but a powerful symbol filled with psychological and allegorical meanings. Noah's flood was not so much an historical event as a symbol of the detrimental effects of the raging passions that brought death and destruction to mankind. And while the Anglican Bishop and member of England's Royal Society John Wilkins (1614–72), who took a literal view of scripture, calculated the exact measurements of the Ark to explain precisely how Noah fed and housed all its occupants, Catholic exegetes took the Ark's measurement figuratively: its breadth of fifty cubits symbolized the fifty days of Pentecost, while its height of thirty cubits represented Jesus's life span.

Peter Harrison contends that the Protestant rejection of the fourfold approach to the biblical text and emphasis on its literal sense historicized the text, drastically circumscribing its meaning: "The Protestant Reformation, by promoting the culture of the literal word, effected a dramatic contraction of the sphere of the sacred, forcibly stripping objects, natural and artificial, of the roles they had once played as bearers of meaning."[41] This kind of historical literalism was an important factor in the demise of the "emblematic" worldview: "the assertion of the primacy of literal reading . . . entailed a new, non-symbolic conception of the nature of things. No longer were objects in the natural world linked to each other by a set of resemblances."[42] Although deemed more perfect and better than any other text, scripture was now part of the historical record. The focus of interpretative efforts centered on the precise meaning the authors intended to communicate. In this respect Protestant scholars followed in the footsteps of their humanist predecessors. Like Humanists they realized that texts had histories and could be understood only in the context of these histories. But there were limits to this approach and clear tensions. For even if the Bible was an historical document, it was still the word of God and its literal interpretation had to be honored. Consequently, while biblical scholarship in the Renaissance and early modern period may historicize, it does not secularize because it never questioned or took a skeptical approach to mystical or miraculous events. As Debora Shuger puts it, scholars combined cultural hermeneutics with theological essentialism. As a result, "The deicidal urges characteristic of modern culture are not emergent . . . [there is] singularly little evidence of encroaching skepticism and secularization."[43] There is therefore a gulf separating the philological and antiquarian scholarship of this period from the Higher Criticism of the late eighteenth and nineteenth centuries.[44]

While this may have been the case for many sixteenth-century biblical scholars, it is not true for all of them, and by the seventeenth century there were a number of individuals who began to question, deny, even ridicule, features of the biblical text. Baruch Spinoza (1638–1712) and Richard Simon (1632–77) were the two most famous or infamous of these.[45] Miracles were an especially sensitive area in this regard, as we shall see. But it was the emphasis on the literal interpretation of biblical passages and the attempt made by scholars to put them in their proper historical context that caused the most problems. Even someone as reluctant to rock the religious boat as Erasmus reveals how dangerous an historical approach could be for determining the correct meaning of biblical passages. When discussing Paul's statement about divorce and remarriage (1 Corinthians 7:39), for example, Erasmus suggests that Paul was addressing a Jewish audience about minor marital problems in cases where reconciliation was still possible. He comments that if Paul were speaking of a "serious conjugal disaster," "he would . . . have interpreted his own writings more humanely for us than we ourselves interpret them." In this instance Erasmus attempts to distinguish a universal injunction from one that is historically contingent.[46] This approach to scripture and biblical history typified the scholarship of both Humanists and antiquarians, whose work contributed to what Grafton describes as a "demythologized" view of antiquity.

Taking the myth out of ancient history was largely the result of the great interest that developed in the material culture of ancient Jews and early Christians. What, for example, did the Urim and Thummim, described as part of the High Priest's vestments, look like, and what were their functions (Ex 28:30)?[47] Did Jews recline during meals like the Romans? What was the nature of Mary's alabaster box, Pilate's atrium, and where exactly was Golgotha located? Every aspect of material culture—clothes, pots and pans, coins, table manners, burial customs, hair styles—piqued the interest of antiquarian scholars, who realized how inadequate patristic scholarship was in regard to these aspects of cultural history.[48] As Sebastian Munster (1489–1552) explained to Henry VIII, newly rediscovered Jewish texts such as the Targums,[49] Midrash,[50] and Talmud,[51] together with Jewish commentaries and Hebrew grammars, offered contemporary scholars an array of new tools with which to understand scripture and the cultures in which it was produced:

> In our era we are assisted by the multitude of books, which we know were unavailable in earlier ages. For St. Jerome himself had no help in interpreting the Old Testament except a naked Bible and an uneducated (and untrustworthy) teacher: no Aramaic translation or Targum, no commentaries, not even a Hebrew grammar—without which many places of Scripture cannot possibly be accurately explained, no matter what some people say.[52]

Grafton singles out the "demythologizing" of antiquity as the greatest accomplishment of the French Protestant scholar Joseph Justus Scaliger (1540–1609):

> . . . Scaliger is moving towards a new view of the ancient world—what might be called a 'demythologized' view of antiquity. It is now absurd to confuse the ramblings of scholiasts about mythological beings with the credible testimony of real historians. It is equally absurd to take seriously the Jewish and early Christian legends about the miraculous perfection of the Septuagint. The ancient world is not different in kind from Scaliger's own. Egypt is no longer for him the land of natural magic and philosophical language. It is simply an old country in the Near East.[53]

Fifty years earlier, Luther had employed this kind of demythologizing to ridicule the Catholic belief in the power of relics. In response to a report that the Cardinal of Mainz was putting some of his newly acquired relics on display, Luther published a spurious catalogue of relics, in which he listed the following items:

1. A good piece of Moses' left horn.
2. 3 flames from the burning bush upon Mr. Sinai.
3. 2 feathers and one egg from the Holy Ghost.
4. An entire corner of that banner with which Christ harrowed Hell.
5. Also, one long hair from the beard of Beelzebub, as it clung to the above.
6. Half a wing from St Gabriel, the Archangel.
7. One full pound of the wind that passed by Samuel in the cave of Mr. Sinai.
8. Two yards of the voice of the trumpet on Mr. Sinai.
9. 30 shakes of Miriam's timbrel heard at the Red Sea.
10. A big heavy chunk of the shout which the children of Israel brought down the walls of Jericho.
11. 3 lovely locks of Absalom's hair, by which his head caught hold of the oak.[54]

The concern with distinguishing fact from fiction was a major aspect of the Reformation as Protestants set themselves the task of separating the original core of Christian beliefs from what they rejected as later Catholic additions, misrepresentations, and distortions.[55] This same preoccupation with distinguishing fact from fiction was a major aspect of the Scientific Revolution, which in important respects paralleled the revolution in religion. Luther's sarcastic attitude toward relics and his reduction of the number of sacraments from seven to two was a product of just such a winnowing process. For example, Luther dismissed the entire sacramental system of confession and penance, along with purgatory and indulgences, because it rested on a mistranslation of the Greek *metanoia*, to repent, as the Latin word *poenitentia*, to do penance. William Tyndale utilized the same kind of philological analysis even earlier, claiming that what Catholics had interpreted as "priest" in reality simply meant "seniors," just as "church" should be translated as "congregation." If the New Testament did not sanction an ordained priesthood and only spoke of congregations and not of a single Church, a reformation of Church organization was in order. This kind of historical and philological approach to scripture was what convinced John Locke and Isaac Newton that the doctrine of the Trinity had no validity because the only reference to it in Scripture was a late interpolation.[56] The French Jansenist historian Jean de Launoy (1603–78), known as the "dénicheur de saints" (the debunker of saints), was a fierce critic of popular superstitions and tenacious in his search for historical accuracy when it came to the credibility of witnesses attesting to saintly miracles.

The realization that the past was different from the present, in other words, an understanding of the concept of history, was a product of biblical and legal scholarship as well as the literary and philological studies of Humanists. Such an understanding of history and historical change was pivotal in the transition from medieval to modern ways of thinking, for it introduced the idea that both law and literature, whether divine or human, was a cultural and historic product and not necessarily valid for all times and places. Debora Shuger draws an amusing analogy between the Humanists' search for authentic forbearers and Freud's "Family Romance." In both cases, the search ends up with frogs instead of princes. As Shuger comments: " . . . [the Humanist] methodology, designed to retrieve the exemplary past from the ravages of time, unearthed alien cultures fixed in time. . . . The Renaissance disciplines struggled with the ancestor who is at once normative and unfamiliar, exemplar and foreigner, origin and alien."[57] A similar encounter with an "alien other" greeted the efforts of French legal scholars seeking to bring local laws into line with Roman law, for what they discovered was how radically different contemporary society was from that presupposed by Roman jurisprudence.

The recognition that law is a cultural construct struck at the very heart of Christianity, for it brought into question both the notion that scripture provided an eternally valid code of divine law and the central Christian doctrine of the atonement. In his treatise *De Jesu Christo servatore* Faustus Socinus (1539–1603), the anti-Trinitarian founder of the Socinians (forerunners of Unitarians), argued that the understanding of the atonement based on the so-called "satisfaction" theory of Anselm of Canterbury (c. 1033–1109) was incompatible with modern notions of individual rights and responsibilities. Socinius objected to the idea that Christ's punishment and death could save others. This belief was based on the notion of vicarious punishment, which, in his view, was unjust. An individual was responsible for his own actions and no one else could or should suffer or pay for the consequences of these actions. Furthermore, because God is the Lord and owner of all things, he could simply have forgiven whatever debt was owed to him; he did not have to insist upon the death of his only-begotten son. The upshot of Socinus's analysis was that Christ's death did not exonerate human beings of their sins or procure some kind of blanket forgiveness. Jesus's death provided an example of faithful and patient suffering meant to inspire people to act virtuously and in such a way as to deserve divine forgiveness. Implicit in Socinus's analysis is the idea that what may be appropriate for one time and place may not hold true for another.

Socinus's emphasis on the responsibility of individuals for their own salvation is one of many examples given throughout this chapter of the dissolution of the holistic worldview characteristic of the Aristotelian-Ptolemaic system and the emergence of a fragmented universe that came in the wake of the Copernican Revolution. Hiram Hayden characterizes the transformation occurring in early modern Europe in just such terms, as "the ultimate desertion of the universal for the particular."[58] No longer a microcosm mirroring a cosmic whole, the human body becomes, like everything else, an assemblage of parts, each one fit for its own particular purpose. Just as the human body was fragmented, so too was the body of traditionally accepted knowledge. In this new universe the grand schemes of encyclopedic knowledge characterizing Gesner's work (along with that of many others) based on the resemblances and correspondences binding every aspect of the physical world into an organic, hierarchical unity collapsed. To illustrate this fragmentation and the "taxonomic confusion" it created, Michel Certeau offers a provocative interpretation of Hieronymous Bosch's *Garden of Earthly Delights*, a triptych whose symbolism and meaning has puzzled art historians for centuries. What Certeau sees in the astonishing scenes painted by Bosch is a pictorial Babel or contemporary *Kunstkammer*, in which real and imaginary objects and entities from all over the world are assembled and depicted with incredible detail and no apparent order or reason. Bosch's painting is a foray into "Ockhamist linguistics," in which words, or in this case images, have no ulterior reference.[59] The triptych presents a parody of an encyclopedia, an image of the world gone mad: "The 'legend' of the Old World . . . is deconstructed by Bosch, who displaces the units of meaning piece by piece; he disturbs, by his hybrids and changes of proportion, the classifying order that was constituted by linking units of meaning together, the same way a sentence is produced by articulating words together."[60] The disproportionality of the objects—huge oysters, fruits, and birds placed alongside tiny humans—defies comprehension. The "grammar" of Bosch's images lack any kind of logic; it is the grammar of "glossolala." The world is indeed illegible as well as unspeakable.

Bosch's Garden offers one of the most intense artistic evocations of the incoherence of early modern life, but he was not alone in his assessment of his times. In her book on seventeenth-century Dutch art, Svetlana Alpers argues that fragmentation becomes a recognized part of seventeenth-century art: "Fragments are prized. . . . No need is felt to pull together, assemble, or in some way resolve individual views into a unified sense of a whole."[61] She cites the rage for "peep-boxes" and still-life paintings of objects from multiple perspectives as an indication of this fascination with multiple and conflicting viewpoints:

> The peep-box, for example, was a construction that also offered various views adding up to make a single world, as do the frequent mirrors or mirroring surfaces; still life obsessively topple containers and peel lemons, or cut pies or open watches to expose multiple aspects to view. One could go on. No single view dominates in the interest of this additive way of piecing together the world.[62]

The peep-box is a perfect image for what had occurred in the centuries after Copernicus. In the same way that the peep-box imposed order on a chaos of conflicting views by artificially establishing the viewer's vantage point, so was order increasingly imposed on recalcitrant humanity from the exclusive vantage point of absolutist monarchs, religious leaders, professionals, and experts. The concept of an organic unity linking individuals to other human beings and to the world at large collapsed in the course of the sixteenth century. What was needed was a new scheme to integrate monadic individuals into a meaningful whole.

CHAPTER 2

MAGIC

On November 7, 1630 Orazio Morandi, Abbot of Santa Prassede and one-time General of the Vallombrosa Order, died in a small cell in the Tor di Nona prison in Rome. The physician writing his death certificate claimed that he saw no sign of poisoning, but the Roman "rumor mill" begged to differ.[1] Four months before his death Morandi was the most revered astrologer in Rome. His precipitous fall from grace reflected the violent, treacherous world of seventeenth-century Italian politics, but his trial records also provide insight into the way religion, science, and magic were inextricably mixed with social and class dynamics in the early modern period as individuals like Morandi jockeyed for power. Morandi was brought to trial because he was an astrologer and dared to predict the imminent death of Pope Urban VIII, who was himself committed to astrology. After Morandi's trial Urban promulgated the severest anti-astrology legislation ever written, and this legislation, coupled with the Pope's heightened awareness of the fractious and downright libertine nature of much clerical culture, all but predetermined the outcome of Galileo's trial. While this facet of the case is obviously important, Morandi's true significance lies elsewhere, in what his career tells us about the dog-eat-dog world of Roman cultural elites and the explosive mixture of religion, magic, and science that characterized the worldview of early modern

---

Portions of this chapter were originally published as Coudert, Allison, "Book Review: Morandi's Last Prophecy and the End of Renaissance Politics by Brendan Dooley." Church History, 2006, 75:190–192. Copyright © American Society of Church History 2006.

Portions of this chapter reprinted from Dewald, Jonathan (editor), *Europe 1450–1789: Encyclopedia of the Early Modern World, 1E*. © 2003 Gale, a part of Cengage Learning, Inc. Reproduction with permission. www.cengage.com/permissions.

Westerners. As Morandi's biographer Brendan Dooley says, "the mystery of Morandi concerns the basic compulsions of advancement in a status-drenched society, and the very nature of knowledge at the origins of science."[2]

Morandi's climb up the social ladder began in Florence, where he joined the circle around Giovanni de' Medici and while in his company became an expert in the occult art of alchemy and a reader of books on the Catholic Index by such authors as Paracelsus, Telesio, and Reuchlin. As a result of Medici patronage, Morandi was appointed Abbot of Santa Passede in 1613. Driven by ambition and his compulsion to gain power and fame, Morandi created a unique library of prohibited scientific, philosophical, theological, and erotic books that he freely lent to those members of the Roman intellectual and social aristocracy who visited his library and attended his elegant soirées. Among the notables attracted were Cardinals on the Congregation of the Index of Forbidden Books as well as Galileo. In addition to lending books, Morandi and his fellow monks provided astrological advice, medicines, charms, and incantations to help people deal with the calamities of seventeenth-century Italian life, characterized as it was by incessant wars, incidents of plague, high mortality rates (especially among children), declining prosperity, increasingly stark divisions between rich and poor, and political instability. It is difficult to exaggerate the centrality of astrology in the lives of people, whether rich or poor, educated or illiterate, living in an age before antibiotics and without social and political institutions (police, firefighters, insurance brokers, etc.) to cushion hardship. People wanted answers, and Morandi was convinced he could provide them by putting astrology on a sound scientific basis.

Morandi was not a charlatan. His goal was to reform astrology and anchor it in a firm basis of observation and experience. In this respect he was in the good company of Girolamo Cardano, Luca Gaurico, Johannes Garcaeus, and Rudolf Goclenius, and many other well-educated intellectuals, who thought that a more experienced-based astrology might not only circumvent ecclesiastical objections but better serve the public.[3] To this end, Morandi planned an astrological encyclopedia of such breadth "that it might have impressed a pioneer scientific organizer like Francis Bacon."[4] This is simply one of innumerable examples of how unstable the very notion of science was at the time. What was true, what false? What disciplines should be deemed "scientific" and what should not? How could one establish a cause or even an effect? For Morandi, as for so many others, astrology and magic were legitimate aspects of natural philosophy, whatever the pronouncements of the Church. After all, the Church had its own magic, and what but pure arbitrariness distinguished that from the magic practiced outside the Church? Morandi's intellectual foundations lay in the Ptolemaic-Aristotelian world, in which man was the microcosm to the greater world's macrocosm. What occurred above went on below. The world was linked in a "Great Chain of Being," in which everything was interconnected in an endless series of correspondences. Symbols were real and connected to things. The eagle and the lion did not simply represent the king; they were the king. Through symbols one could "tap into the very viscera of things" and change the world.[5]

Morandi was arrested after Urban was informed that he was the person who had predicted his death. But this created an impossible situation because it soon became evident that the case involved a significant segment of Roman high society, and exposing the libertine culture of Morandi's monastery would be hugely embarrassing to the Church. In these circumstances, Morandi's death could not have been more convenient.

Morandi was one of history's innumerable casualties. Like the vast majority of people before and after him, he was all but obliterated from the historical record, but his life story adds appreciably to a more nuanced and complex understanding of the social, religious, and political forces that shaped the Scientific Revolution and with it the emergence of the modern world. At the time of Morandi's death it was not at all clear what science was and who would be remembered for their innovations and who ignored or even vilified for retrograde thinking. In short, no one knew exactly how to differentiate between religion, magic, and science, or, if they claimed they did, their conclusions had the polemical purpose of declaring their opinions correct while the opinions of everyone who disagreed were not only wrong but demonic. Thus "magic," like the word "superstition," became a pejorative term hurled with abandon during a period of acrimonious religious conflict.

Does this mean that the concept of magic is essentially meaningless and should be rejected as some scholars suggest?[6] This suggestion makes it all the more interesting, as Randall Styers argues, that magic has been a central theme in the literature of the social sciences and religious studies from their inception. Styers attributes the centrality of a concept that is so elastic and imprecise to the fact that "[d]ebates over magic provide an extraordinarily rich ground for exploring the nature of modernity, its values, and its limits."[7] Because definitions by their nature involve distinguishing one thing by opposing it to something else, magic has proven useful in separating what is supposedly "modern" from what is "non-modern." In this case, modernity has been defined, admittedly by moderns, as "rational," "progressive," and "scientific," everything in fact that magic is considered not to be. Frazer's famous definition of magic as pseudo-science or the "bastard sister of science" fits this paradigm: "In short, magic is a spurious system of natural law as well as a fallacious guide of conduct; it is a false science as well as an abortive art."[8] Magic has also proven useful for defining religion and science. Starting with Durkheim and Mauss and reaching a crescendo in Voeglin and Bateson, magic has been vilified as a corrupt form of religion because it presupposes that human beings can change and transform the world. As Styers puts it so well, "Numerous scholars have argued that magic involves willful, assertive action—a failure to submit to the inexorable divine and natural order."[9] What this evaluation of magic so flagrantly obscures is the fact that determining what is "the inexorable divine and natural order" has been the role of religious leaders for millennia, and transforming the human and natural order has been and still is the goal of most religions and religious leaders. To suggest that religion is uncontaminated by any idea of seeking power is to ignore the material effects of religious ideas and the power that religious institutions have wielded and still wield over individuals, governments, and nations. The same is true of juxtaposing "good" science and "bad magic." Good science is supposedly objective, neutral, and value-free, while magic is private, self-serving, and anti-social. The idea that the magician oversteps the boundaries between the divine and human while the truly religious person as well as the exemplary scientist are suitably humble and submissive in the face of a higher reality is to draw a fictitious picture of religion, magic, and science. Yes, the magician blurred the lines between the human and the divine, but who drew this line in the first place and whose interests does such a line serve? Even more importantly, who cannot think of religious leaders and scientists who have either willingly assumed the role of God or been charged with assuming such a role? These are questions to keep in mind throughout this book because they will repeatedly come up. The solution is neither to accept these false dichotomies nor to throw out the categories of religion, magic, and science altogether, but to gain an understanding of how they work in conjunction with each other, meaning different things to different people at different times. Magic, as we shall see in this chapter, meant the highest, most laudable, and holy kind of knowledge possible to Cornelius Agrippa and many of his fellow magicians. But to Catholic inquisitors and Protestant theologians magic was a dangerous and illegitimate usurpation of demonic powers and practices. Why and how individuals could think so differently about the supposedly same thing is of crucial importance if we are to understand the early modern period, not to mention all periods including our own.

Abby Warburg and the scholars at the Warburg Institute he founded were in the forefront of those fully aware of the importance of magic and esoteric forms of thought in understanding the art and culture of every period. This was particularly true in the case of Renaissance Humanists like Pico and Ficino (who studied esoteric sources in the form of Kabbalistic, Hermetic, and ancient magical and astrological texts), not to mention artists, poets, and writers who made use of magical and esoteric themes in their work.[10] Historians of science were slower off the mark but for the most part now recognize that magic, especially "natural" magic, had something to do with the development of new theories of nature and new ideas about the value of experimentation in the emergence of modern science, although they differ as to exactly what that was. The debate over the exact meaning of "natural" in "natural magic" was part of the more extensive debate about nature and its relation to humanity and the divine. What had previously seemed like fixed and unalterable categories in the ancient and medieval world were so no longer. What was natural, human, or divine? Charles Taylor and Ariel Glucklich highlight the role that the suppression of magical thinking played in the emergence of modern forms of subjectivity and identity, which stress inwardness and create stronger boundaries between the self and others.[11] It takes an early modern historian like Lyndal Roper, however, to spell out how costly the development of individualism was in regard to the social control and regimentation exerted on individuals in the early modern period.[12] As a result of insights like hers, the connection between the suppression of magic and the centralization of political power has become another subject highlighted by recent research into early modern history.[13] Historians and particularly postmodernists have also been assiduous in describing how theories that equate magic with primitive and anti-social thinking were instrumental in supporting imperialism and colonialism. Few Americans know, for example, that Rudyard Kipling's poem "The White Man's Burden" (1899) was written as an appeal to the United States to take over rule of the Philippines, recently won in the Spanish-American War. As McKinley's viceroy in the Philippines, William Howard Taft, put it, we had to help "our little brown brothers," who in Kipling's memorable words were "fluttered folk and wild—Your new-caught, sullen peoples, Half-devil and half-child."[14]

As the middle term sandwiched between religion and science, magic has some of the characteristics of both. In the early modern period magic was, as Keith Thomas tried to convince the anthropologist Hilda Geertz, something that was very real to most people and not simply a pejorative term, although it was that as well.[15] There were actual men like Cornelius Agrippa, Gianbattista della Porta, and Johannes Trithemius who proudly described themselves as "magi" (the plural of "magus," Latin for magician). There were even networks of such self-professed magi.[16] As Morandi's career reveals, whether Catholic or Protestant, pope or peasant, the overwhelming majority of people lived in a miraculous and magical universe infused with the supernatural. They inhabited a "moralized universe," to use Robert Scribner's apt phrase, in which human actions could provoke supernatural intervention. While some people resigned themselves to this, most did not and made great effort to anticipate, deflect, or turn supernatural forces to their advantage. Given this state of affairs, it is fair to say that far from "desacralizing" or "disenchanting" the world, at least in the short term, Calvin and Luther intensified fear of the supernatural to an unprecedented degree by emphasizing the cosmic struggle between the divine and the diabolical while placing man at the mercy of both. Citing Paul and the Gospel of John, Luther describes the devil as the "prince and God" of this world: "we be strangers in this world, whereof he is the prince and God. Therefore the bread which we eat, the drink which we drink, the garments which we wear, yea, the air, and whatsoever we live by in the flesh is under his dominion."[17] Luther saw devils everywhere, and in one famous incident threw an inkpot at the devil when he dared to interrupt him as he translated scripture: "Doctor Luther sat at the Wartburg translating the Bible. The Devil did not like this and wanted to disturb the sacred work, but when he tried to tempt him, Luther grabbed the ink pot from which he was writing, and threw it at the Evil One's head. Still today they show the room and the chair where Luther was sitting, and the spot on the wall made by the flying ink."[18] Calvin attributed the same formidable power to the devil. As he writes in *The Institutes*:

> We have been forewarned that an enemy relentlessly threatens us, an enemy who is the very embodiment of rash boldness, or military prowess, of crafty wiles, of untiring zeal and haste, of every conceivable weapon and of skill in the science of warfare. . . . Scripture makes known that there are not one, nor two, nor a few foes, but great armies, which wage war against us. . . . [19]

Given the power attributed to the devil by Luther and Calvin it is entirely understandable that Protestants engaged in many of the same ritual practices employed by Catholics to protect themselves against the myriad malign forces threatening them. In fact, by abolishing the magical practices of the medieval Church, Protestantism actually promoted popular reliance on cunning men and women, astrologers, and magicians. Consequently, the thesis made famous by Max Weber in *The Protestant Ethic and the Spirit of Capitalism* that Protestantism "disenchanted" the world needs to be modified. While in the long term Protestantism did contribute to this disenchantment as well as to the secularization that accompanied it, this process took much longer than Weber imagined. Protestantism did not arrive fully formed in thought or practice. Gerhard Strauss documents the despair of many Protestant reformers, and this includes Luther, who felt that the Reformation had failed to root out the idolatry and superstition they associated with Catholicism.[20] Nevertheless, over the long term the Protestant rejection of the Catholic sacramental system initiated a line of questioning that contributed to the crisis of skepticism already brewing as a result of humanist scholarship and the information explosion. For by denying that rituals had any effect beyond a symbolic one, Protestants undermined the need for an institutional church or a separate caste of priests endowed with the supernatural power to supervise and perform salvific rites. Euen Cameron makes this point as a corrective to those social historians who have been most critical of Weber's thesis, arguing that Protestantism did little to disenchant the world. As Cameron says, "Some social historians who confidently assert that there was nothing demystifying or 'disenchanting' about Protestantism should take into account that the core theology of the Reformation faith was *in its very essence* a process of demystification."[21] Cameron readily admits, however, that the way this basic Protestant message was received at the popular level is a wholly different matter. What seems clear is that different groups of Protestants received the message differently and that wherever the process of disenchantment did occur, it didn't happen overnight and in most places never happened completely.[22] It is also clear that Catholics experienced disenchantment as Catholic theologians tried to separate Church ritual from magical ritual with varying degrees of success. For this reason this chapter along with the next two will investigate the reasons why early modern people, whatever their denomination, continued to believe in magic, miracles, and witchcraft. But these chapters will also discuss the ways in which magical beliefs were undermined by those who had drunk deeply from skeptical wells and questioned the idea that supernatural beings could intervene in the natural world. These skeptics, while a definite minority, raised enough questions to make those who defended the existence of magic, miracles, and witchcraft go on record in an attempt to justify their beliefs. This debate, which involved religion as much as science or natural philosophy, was a key factor in what later led to the disenchantment and secularization of the world.

As we have seen, the devil assumed a central position in Protestant thought, one he never quite achieved among Catholics. For while Luther, Calvin, and other Protestant reformers emphasized the fact that life was one long struggle against the demonic "other," they removed the support of priests, saints, and especially the Virgin Mary enjoyed by Catholics to help them get through tough times. The ascendency of the devil in Protestant thought was reflected in the huge popularity of a new genre of popular literature, the devil book. By conservative estimate there were approximately 100,000 individual copies of devil books on the German market alone during the 1560s, a truly enormous number if one considers the low level of literacy at the time. In their most common form devil books singled out a particular vice—smoking, drinking, dancing, gambling, or swearing—and showed how devilish it was. We can see this, for example, in the title page of an English devil book that castigates the devilish practice of smoking tobacco. Here we see some kind of hairy beast standing on two legs with a long tail as he smokes a pipe. Devil books also attacked specific religious and political groups at a time when religion and politics were thoroughly entwined, as one can see in another frontispiece that puts "round heads" or Puritans in the devil's camp.[23] The sale

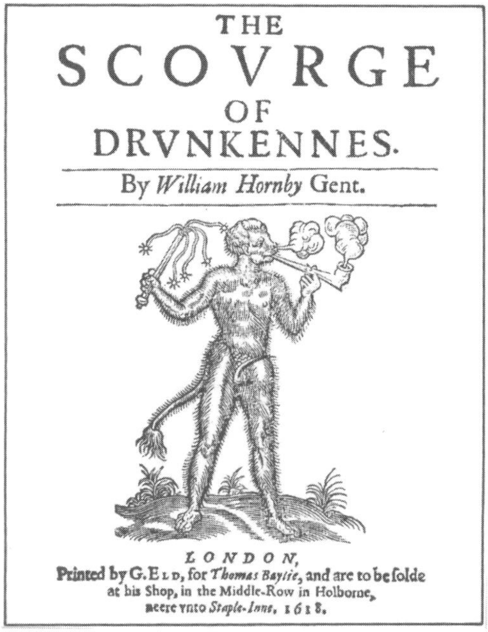
THE
SCOVRGE
OF
DRVNKENNES.
By *William Hornby* Gent.

LONDON,
Printed by G.Eld, for *Thomas Baylie*, and are to be solde
at his Shop, in the Middle-Row in Holborne,
neere vnto *Staple-Inne*. 1618.

Devil of tobacco "drinking." (From William Hornby's *The Scourge of Drunkeness*. London, 1618. From Ernst and Johanna Lehner, *Picture Book of Devils, Demons and Witchcraft*. New York: Dover Publications, Inc., 1971, 167.)

of devil books was forbidden in Catholic countries, and although they were smuggled in and read by Catholics, devil books were a characteristically Protestant form of literature.[24]

The difference between Protestant and Catholic attitudes toward the idea of a demonic pact offers another example of the heightened fear of the devil among Protestants. The prominence of covenant theology in Protestant thought had its dark side. If a covenant could be made between man and God, a diabolical one was equally possible. Such a pact did not even require a formal declaration, simply the intention to sin. A comparison of the Catholic story of Theophilius, a monk, who made a pact with the devil, and that of Dr. Faustus reveals that it was virtually impossible for Protestants, but not Catholics, to renounce a pact once made. The Catholic Church told people exactly how to cheat the devil by calling on the Virgin or the saints. Theophilius followed instructions, and the Virgin saved him. Faustus, for whom recourse to the Virgin was out of the questions, suffered a horrific death and went to hell.[25]

THE
DEVIL TURN'D
ROUND-HEAD:
OR,
PLVTO become a BROWNIST.

Being a juſt compariſon, how the Devil is become a *Round-Head?* In what manner, and how zealouſly (like them) he is affected with the moving of the Spirit.

With the holy Siſters deſire of Copulation (if he would ſeem Holy, Sincere, and Pure) were it with the Devill himſelf.

As alſo, the Amſterdammian definition of a Familiſt.

The Round-Head Devil. (From John Taylor, *The Devil Turn'd Round-Head.* London, 1642. From Ernst and Johanna Lehner, *Picture Book of Devils, Demons and Witchcraft.* New York: Dover Publications, Inc., 1971, 170.)

For early modern men and women devils and demons were not the only sources of evil and misfortune; witches, fairies, elves, and ghosts were just as potentially dangerous. To avoid all these baneful creatures with their malevolent influences and to attract positive ones, rituals accompanied every aspect of daily life from the production of food and the care of animals to the building of houses and the birth and raising of children. Despite the concerted attack on Catholic "superstition" and "magic" by Protestant theologians, Protestants were just as engaged in these kinds of prophylactic magical rituals as Catholics. They used the Bible as a magic object to test for witchcraft and divine the future. They even carried it on their persons as a protective amulet. They consecrated church foundation stones, pulpits, fonts, organs, cemeteries, and pilgrimage sites. And if by chance Protestant clergymen would not supply them with the consecrated items they believed they needed, they were quite willing to resort to Catholics. Although some Protestants rejected astrology, the majority did not. After all, the rallying cry of Protestants was "sola scriptura," and scripture legitimized astrology: "there will be signs in the sun, the moon and the stars" (Luke 21:25). Thus magical and miraculous events were accepted aspects of everyday life. Some scholars continue to distinguish magic from religion on the grounds that magic attempts to manipulate supernatural forces while religion only supplicates them,[26] but this distinction is untenable.[27] Religion, in whatever form it came in the early modern period, was as much functional as devotional. What ordinary people wanted were results—good fortune and good health if possible, survival at the very least—and to procure these results they were willing to try anything that seemed promising. They did not just beg or petition for results; they demanded them and took steps to ensure their demands would be met.

Magic provides an example of how pre-Copernican beliefs about the universe persisted after the Copernican Revolution, which should have and to some extend did undermine the basic assumptions on which magical beliefs and practices rested. Magic, as we have seen, was based on a view of the world as an integral whole composed of interacting spiritual and material forces that human beings can manipulate for good or evil purposes. In this sense, magic was predicated on the emblematic worldview described in chapter one that provided the rationale for Camillo's theater. Magic encompassed a wide range of activities such as astrology, alchemy, medicine, divination, necromancy, and conjuring. While this definition holds true for magic over the millennia, only during the early modern period was "black" magic equated with demonic witchcraft and made into a serious criminal offense. At the same time that black magic was demonized there was a growing interest in and respect for "natural" or "spiritual" magic, which began in the twelfth century, reaching its apogee during the Renaissance and early modern period. This form of magic contributed to the Scientific Revolution, while the intensified fear of demonic magic was an important factor in the witch hunts, which, in their own paradoxical way, also contributed to the Scientific Revolution.

Much of early modern magic represented a continuation of traditions and practices that developed in the medieval period from a synthesis of classical, Jewish, Islamic, and Christian concepts of magic with the incorporation of indigenous Celtic, Germanic, Scandinavian, and

Slavic traditions as these groups were converted to Christianity. It is difficult—though in some cases possible—to separate these various strands because they were so thoroughly mixed with Christian elements. Christianity shared many assumptions that were basic to a magical worldview, which is why it has proven so challenging to make a clear distinction between the two. Foremost among these assumptions was the idea of a universe that was alive and vital and divided into three levels, the super-celestial, celestial, and terrestrial. Each of these levels was intimately linked to the others through a series of correspondences, sympathies, and antipathies that might be hidden (occult) but that were regular, rational, and discoverable. This was the basis for the idea of "The Great Chain of Being" described in chapter one. Christianity and magic also agreed about the existence of invisible, spiritual entities (angels, demons, devils) who interacted with humans in innumerable ways, including sexually.[28] Christianity and magic both emphasized the power and efficacy of words, a belief that was intensified by the Christian reliance on the spoken and written word and by the notion of Christ as the incarnate word of God. Many magical prayers and formulas were simply adaptations of Christian formulations.[29] A further link between Christianity and magic was the belief that hidden powers and virtues existed in natural objects (amulets, talismans, relics, holy water, the sign of the cross, the Eucharist, church bells), which could be tapped for human use. Given these similarities, one can agree with Stuart Clark that "[a]cross Europe, throughout the centuries . . . magic often seems indistinguishable from religion."[30]

On a popular level magic was practiced extensively to deal with difficult situations from childbirth and childcare, to animal husbandry, sickness, misfortune, lost or stolen objects, divination, business affairs, traveling, falling in or out of love, counteracting witchcraft, and even such mundane activities as shutting windows at night.[31] Magical remedies, rituals, and formulas can be found in necromancer's manuals, medical textbooks, scientific texts, the lives of saints, and courtly romances. While astrology was a recognized part of academic medicine, magical healing was reserved primarily for diseases considered "unnatural," such as madness, possession, and nightmares, or those diseases whose causes were unknown and consequently attributed to the evil machinations of sorcerers, witches, demons, elves, and the like as in the case of sudden strokes, heart attacks, and seizures. In these cases, magicians and healers patterned their actions after those of Jesus and the saints, conjuring spiritual forces by ritual actions or reciting prayers, blessings, and exorcisms. They also used amulets, talismans, relics, the sign of the cross, holy water, and nostrums made variously from herbs, animal parts, stones, or gems. Next to healing the most popular form of magic was divination, a practice emphatically rejected by Christian authorities. Charts and manuals existed for reading signs about the future in the sky or in animals, plants, parts of the human body, as well as in dreams. Love magic was used both to seduce and to cause impotency, a common theme in both courtly romances and inquisitors' manuals.[32]

Language and symbols played an important part in the debate over magic that became particularly heated from the late fifteenth through the seventeenth centuries. This debate, in turn, was part of the larger issue of how to distinguish "natural" or "spiritual" magic from illicit and "demonic" magic and so carve out a realm for the legitimate practice of natural philosophy or science.[33] On one side of the debate were the Neoplatonists, Hermeticists, and Kabbalists, who considered magic a natural force in an animate universe and advocated their own brands of "good" natural magic as fully legitimate and scientific On the other side were those Catholics and Protestants for whom magic was a diabolical perversion of their faith. A central issue in this debate was whether words and images had magical power and, if they did, whether this power was natural or demonic and its use licit or illicit. Those in favor of natural magic generally agreed that images and words (especially divine names and the words in a holy language) were powerful forces and could be legitimately used to produce good magical effects. Their opponents either insisted their use was diabolical or denied that they had any force at all. Underlying this last argument was the assumption that language and symbols were conventional and not "real" representations of things; therefore they were powerless to produce magical effects.

Marsilio Ficino (1433–99) was one of the earliest, most famous, and influential of Renaissance natural magicians to discuss the magical power of images and sounds. His ideas were embellished by innumerable later writers. He developed a form of spiritual and subjective magic to attract beneficial celestial forces into the soul of the operator.[34] An essential element in Ficino's magic was his conviction that words represent the natures of things. Like so many others, Ficino accepted Plato's premise that the words in a perfect language would indicate the nature of things but ignored his conclusion that such a language could not exist.[35] Ficino argued that words can become powerful forces when uttered, particularly when set to music, for song is "warm air, even breathing, and in a measure living, made up of articulated limbs, like an animal, not only bearing movement and emotion, but even signification, like a mind, so that it can be said to be, as it were a kind of aerial and rational animal."[36] Although Ficino tried to confine his spiritual magic to subjective effects, there is the clear implication that the effects could be transitive, an implication Cornelius Agrippa and other less circumspect magicians openly awowed.

Like popular magic, "natural" or "spiritual" magic was also concerned with issues of healing, protection, and divination, but there was an emphasis on the intellectual and moral character of the natural magician lacking in popular forms of magic. This is one more example of how tenuous the distinction between religion and magic is. Because to successfully manipulate the world, the magus had to possess a clear understanding of the inherent properties of things, and this involved an appreciation of the highest mysteries. Agrippa provides a good summation of the religious caste and sacred nature of this kind of magic:

> Magic is a faculty of wonderful power, full of most high mysteries. It contains the most profound contemplation of things, which are most secret, together with their nature, power, quality, substance and virtues, and the knowledge of the whole of nature. It instructs us in the way things differ and agree with each other and thus it produces wonderful effects by applying the virtues of one thing to another and thus uniting them. It also joins and knits firmly together compatible interior objects by means of the powers and virtues of superior bodies. This is the most perfect and principal branch of knowledge, a sacred and more lofty kind of philosophy, and the most absolute perfection of every most excellent philosophy.[37]

To acquire this most excellent philosophy required the highest degree of intelligence and moral probity combined with a deep knowledge of theology:

> So whoever wishes to study this faculty must be skilled in natural philosophy in which is to be found the qualities of things and the

> hidden properties of everything which exists. He must also be expert in mathematics, and in the aspects and figures of the stars, upon which depends the sublime virtue and property of everything; and in the theology in which are manifested those immaterial substances which regulate and administer all things. Without these, he cannot possibly be able to understand the rationality of magic.[38]

In these passages Agrippa virtually erases the distinction between natural philosophy and magic, and he adds another important feature of natural magic, its association with mathematics. Throughout the Middle Ages and Renaissance technology, mathematics, and magic were closely allied, especially when it came to the mechanical contrivances producing marvelous effects that were so popular in public entertainments and courtly masques and fêtes. Technology, mathematics, and magic continued to be linked during the Scientific Revolution, although technology and mathematics gradually shed their association with magic and came into their own as bona fide scientific subjects.[39]

Agrippa insists that magical knowledge must be grounded in moral excellence and faith in God and that in large part such knowledge comes through divine inspiration.

> When we, by the remembrance of [divine science's] majesty always busied in divine studies, do every moment contemplate divine things, by a sage and diligent inquisition and by all the degrees of the creatures ascending even to the Archetype himself, we do draw from him the infallible virtue of all things; which those that neglect, trusting only to natural and worldly things, are wont often to be confounded by diverse errors and fallacies, and very oft to be deceived by evil spirits. But the understanding of divine things purgeth the mind from errors, and rendereth it divine, giveth infallible power to our works, and driveth far the deceits and obstacles of all evil spirits, and together subjects them to our command.[40]

From this statement, it is hard to deny that for Agrippa natural magic is characterized by its "intense religiosity and sense of piety," but it is a piety that exalts human nature rather than debasing it.[41] Fallen man has the power within himself to purge his mind of errors and become divine. This was the theme in Pico's "Oration," and it was and is a theme in so-called "gnostic" literature from the second-century Gospel of Thomas until today. As Agrippa says, "for by how much the more we have relinquished the animal and the human life, by so much the more we live like angels, and God, to which being conjoined, and brought into a better condition, we have power over all things, ruling over all."[42] Agrippa's emphasis on moral rectitude and divine illumination was not unique. Many other proponents of natural magic concurred. This was a common theme among alchemists and one that Ben Jonson used to great comic effect in his play *The Alchemist*, when the conniving alchemist, Subtle, points to the immorality and greed of his gullible client, Sir Epicure Mammon, as an excuse for his failure to transmute base metal into gold. The goal of Agrippa's natural magic, like Ficino's, was to attract beneficial divine and spiritual forces into the soul of the magus, who would then use them to manipulate matter. Agrippa's detractors, however, disagreed, insisting that all magic was diabolical and that as the worst of magicians Agrippa deserved and got a fate worse than death. This was the opinion of James Sanford:

> . . . in the ende his wicked knowledge was the cause of his miserable deathe: for as John Manlius a Germaine writer doth recorde, when he was at the point of death, he called to him a dog which went about with him, and spake to him with these wordes, *Abia me perdita bestia, quae me perdidisti*: that is, Depart from mee thou wicked beast which hath destroyed me. So forthwith the dog departing from him, caste himselfe headlong into a river. This dog was without doubt a Divel of Hell.[43]

Owners of black poodles may be interested to know that was the breed and color of Agrippa's questionable canine companion.

Scholars have differed wildly in their evaluation of Agrippa. Lynn Thorndike claimed that he had simply produced an unoriginal compilation of other writers' ideas with no theory or focus of his own. The apparent contradiction between Agrippa's fervent espousal of magic in his *Occult Philosophy* and his later apparently equally fervent rejection of it in his *On the Incertitude and Vanity of the Arts and Sciences* has led to the conclusion that Agrippa was simply incoherent. But his emphasis on faith as a prerequisite to knowledge has solved the problem of this apparent incoherence for Christopher Lehrich. Lehrich makes the convincing argument that the two works are consistent because in the latter Agrippa does not claim that all knowledge is worthless, only knowledge that is not firmly grounded in faith.

If Agrippa receives a better press these days, what about natural magic in general? As suggested earlier, most scholars are presently prepared to admit that natural magic, along with esoteric thought of various kinds, played a role in the emergence of modern science. This is an issue taken up in chapters seven and eight, but for the moment what can be said is that natural magicians like Agrippa and della Porta advocated experimentation and were interested in all kinds of technical inventions, recipes, and processes dealing with agriculture, medicine, alchemy, and the many subfields grouped under the rubric of natural philosophy. They were also willing to test experimentally the supposed boundaries between the natural and supernatural. Della Porta, for example, experimented on an old woman who claimed to have successfully used an ointment enabling witches to fly to demonic sabbaths. After she fell into a trance, he beat her so fiercely that she would see the bruises when she woke up and realize she had gone nowhere. She still claimed to have flown to the witches' Sabbath, but he convinced her that it had been a dream induced by the narcotic in the ointment.[44] Della Porta's work on opticks, mirrors, rainbows, light, water vapor, and a host of other natural phenomena influenced many others who took up and profited from his suggestions. Even Marin Mersenne, who was highly critical of magic, wanted to purge della Porta's work of errors, not ban it.

The work of natural magicians is hard to classify. They offer lofty claims of conversing with angels but at the same time present mundane recipes for boot polish or hair dye. Such is the case with della Porta's *Magia naturalis* (enlarged edition 1589), an extremely popular manual on natural magic that describes procedures for such diverse things as transmuting metals, producing exotic plants and animals through grafting and cross-breeding, cutting, conserving, and cooking meat, staving off baldness, eliminating wrinkles, and engendering beautiful children. Agrippa's *De Occulta Philosophia* contains the same mix of the sublime and the commonplace. While he describes a magus talking to angels and creating "living" statues and marvelous mechanical devices, he also advises the reader that the hearts or genitals of animals are effective ingredients in love potions.

It is not always easy to distinguish "natural" from "demonic" magic, for one man's "natural" magic could clearly be "demonic" in the eyes of another. Nevertheless, necromancy and black magic were an established part of medieval magic and continued to be practiced throughout the early modern period, although with growing disrepute. The *Picatrix*, derived from an Arabic source that mixed spiritual and demonic magic with astrology, was widely influential.[45] The increasing fear of demonic magic on the part of religious and secular authorities began around 1400, only developing into a full-blown panic in the early modern period. No longer seen as a body of superstitious and largely illusory practices that could be eradicated through a combination of missionary activity and the counter use of Christian ritual—a view characteristic of the Middle Ages—magic and magicians came to be viewed as a demonic fifth column threatening the very existence of Christian civilization. This negative view of magic was reinforced by the Protestant attack on Catholic sacraments, rituals, and miracles as demonic. For the most part, however, Catholic and Protestant authorities distinguished between "popular" magic, whose practitioners were prosecuted as witches and sorcerers in league with the devil, and "learned" or "spiritual" magic, which was generally tolerated and widely practiced at European courts because of its promise of wealth and prestige in addition to its sheer entertainment value.[46] Even when tolerated, magicians inspired ambivalent attitudes because beneficent "white" magic might easily be perverted into "black" magic. For this reason two of the foremost demonologists of the sixteenth century, Jean Bodin and Martin Del Rio, condemned all magic as demonic.

The increased concern with demonology and witchcraft in the early modern period has been attributed to the religious conflicts stirred up by the Reformation and Counter-Reformation.[47] Recent research has shown, however, that it was not religious conflict per se that encouraged witch hunts but the new age of "confessionalism" accompanying these conflicts and heightening religious fervor and the concern with eradicating religious deviance.[48] In addition, the increased fear of magic and sorcery can be seen as a response to the collapse of the Aristotelian-Ptolemaic worldview and the uncertainty this collapse generated. A new world order was called for that reaffirmed the goodness of God and the reality of divine providence. This was supplied in large part by early modern demonologists and witch theorists. Neither irrational nor unscientific, they deployed all the resources available from natural philosophy and theology to vindicate divine justice and the truth of the Bible.

Witchcraft theory was a kind of "theological damage control," to quote Walter Stephens, that let God off the hook of seeming injustice by attributing evil and misfortune to the activities of men and women in league with the devil.[49] Exactly how witchcraft accusations and convictions helped to create a new world order is the subject of a later chapter. Before broaching that complex subject a word must be said about the role of miracles in the early modern period.

CHAPTER 3

MIRACLES

Just as it used to be the accepted wisdom that Protestants rejected magic, so too did scholars believe Protestants rejected the possibility of miracles, at least those said to have occurred after the first few Christian centuries. While many Protestant pastors and theologians attacked the miracles flaunted by the Catholic Church as a sign of its superiority and legitimacy and ridiculed the saints supposedly responsible for them—Luther went so far as to suggest they were satanic delusions—they were unable to eradicate belief in miracles among their parishioners. Although reports of miracles declined in some areas, especially Germany in the period between 1520 and 1570, such reports continued because they satisfied the needs of many Protestants to have proof of God's active concern with the world. As Philip Soergel has shown, "a lush undergrowth of stories about miracles . . . continued to satisfy the desires of Lutherans, as well as those of Catholics and Calvinists, for cases of divine intervention."[1] A very good indication of just how widespread magic and miraculous events continued to be in the early modern period is contained in the visitation reports of Protestant reformers, who made a concerted effort to question people about their religious beliefs. Strauss describes the disappointment of the reformers when they realized that their hopes for a renewal of religious and moral life through mass indoctrination of the Gospel had been received for the most part with "utter indifference" and a continued belief in magic and miracles:

> . . . there can be little doubt that magic cults held the trust and engaged the interest of the majority of the populace . . . and that the operative religion of country folk, and perhaps of many city-dwellers

> as well, had much less to do with the doctrine of established Christianity than with the spells, chants, signs and paraphernalia of ancient magic lore and wizardry, the cult of which flourished unaffected by the imposition of new or old denominational creed.[2]

Given the vociferous condemnation of saints by Protestant reformers, it may be startling to realize that from the earliest days of the Reformation, as well as later, Luther was himself portrayed in images very like those of medieval saints.[3] In one case, a portrait of Luther provided Protestant soldiers with their own sacred and magically powerful icon during the Thirty Years' War. While Catholic soldiers touted a crucifix that withstood the attempts made by Swedish troops to burn it, the Swedes claimed they had an engraving of Luther that was equally incombustible.[4]

In those Protestant areas where miracles and the cult of saints were condemned most fiercely, reports of prodigies, omens, signs, wonders, and marvels multiplied exponentially. In this regard, Protestants were no different from Catholics in seeking signs that would help them anticipate and cope with an unpredictable future or explain what had already occurred.[5] The only difference, according to Soergel, was that Protestants "vastly outproduced their Catholic Counterparts" in their creation and dissemination of this material and this continued for a good one-hundred-and-fifty years after the onset of the Reformation.[6] Monstrous births were an especially rich source of speculation as to the sin that caused them. John Winthrop (1588–1649), the Puritan leader and governor of Massachusetts, was immensely gratified to find that his condemnation of Anne Hutchinson (1591–1643) for her criticism of the established Church and its male leadership was vindicated by her bringing forth "not one . . . but . . . 30 monstrous births or thereabout, at once." He saw the hand of God in this event: "See how the wisdom of God fitted this judgment to her sin in every way, for look as she had vented mis-shapen opinions, so she must bring forth deformed monsters."[7] The satisfaction, even glee, with which individuals interpreted the misfortunes of their neighbors as signs of God's displeasure is somewhat surprising for a religion supposedly based on love. But the attempt to discern divine messages from nature was simply too tempting at a time when there was no clear concept of cause and effect in many areas of human life, especially medicine, and when disaster could strike without warning. As a consequence, any event however insignificant or terrible could reveal the hand of God.

Ralph Josselin (1616–83), the Vicar of Earls Colne in Essex famous for his diary, was stung on his nose by a bee. The fact that his nose failed to swell was, for him, a clear sign of divine providence. After the death of his dearly beloved eight-year-old daughter Mary, he wondered if she had died because he was too fond of card playing.[8] His contemporary, the Puritan theologian Richard Baxter (1615–91), believed that the death of his wife was a warning from God to remind him of his sins.[9] When the Salem judge Samuel Sewell's (1652–1730) glass of spirits "fell down and broke all in slivers," he wrote in his diary, "I said twas a lively Emblem of our Fragility and Mortality."[10] As Soergel has put it so well, for early modern men and women "the earth was like a vast book in which God was continually writing strange and terrifying texts."[11] While Soergel writes about Germany, this was true for Protestants across the continent and in America.[12]

The Protestant view of miracles was therefore paradoxical. They were unnecessary and even harmful, if not fraudulent, because all that was required for salvation was faith, and faith had to be freely given. Yet at the same time miracles offered proof that God was active in nature and concerned enough for his worshippers to offer them clues about what they should do. Thus despite the claim made by some Protestant theologians that miracles had ceased, many Protestants continued to believe in them, and the number of people declaring themselves miracle workers actually increased in England during the early modern period.[13] George Fox (1624–91), the founder of the Quakers, kept a "miracle book," in which he claimed to have raised the dead.[14] Baptists healed according to biblical models, and Charles II touched more subjects for the King's Evil than any of his predecessors, some 90,000.[15] Jane Shaw notes an upsurge in miracles in England in the 1690s, attributing this to the fact that the war against France was going badly and to the apocalyptic and millennial expectations inspired by the approaching new century.[16] She describes what she calls a "perfectly Protestant miracle" in the case of a thirteen-year-old Huguenot refugee by the name of Marie Maillard, who was living in London with Mme. Renée de Laulan. Marie had been lame from birth as the result of a tumor, and eminent Huguenot surgeons had declared her incurable. On November 26, 1693, however, she was instantly healed while reading the second chapter of the Gospel of Mark, which describes the healing of the man with palsy. Crowds came to visit Marie. She was summoned to appear before Lady Sutherland, the Lord Mayor of London, and Sir William Ashurst. Four surgeons were sent by Queen Mary and at least three by the Bishops of Worcester, Salisbury, and London to investigate the validity of the miraculous cure. An anonymous ballad, "The Happy Damsel: or, a Miracle of God's Mercy" was published, along with a number of accounts of the event.[17]

Marie's miraculous cure fit the bill of a "perfectly Protestant miracle" because there were no intermediary figures involved; the miracle occurred as a direct result of reading the Bible; and it happened through God's direct intervention without any request from Marie herself. This exemplified the Protestant emphasis on God's total power over the universe and the Bible as the preeminent scriptural text.[18] Miracles connected to Bible reading were quintessentially Protestant because Bible reading was a key form of Protestant worship before toleration was accepted in England and when public worship by Dissenters was forbidden. Such reading was the central activity for Huguenots in France, especially after the Revocation of the Edit of Nantes in 1685, and for Protestants in any areas of Europe where the practice of their religion was criminalized. Marie's cure was also seen by many Protestants as a remedy for the atheism they believed was plaguing society. Deuel Pead, minister of St. James in Clerkenwell and chaplain to the Duke of Newcastle, described Marie's cure as

> a wonderful work, a signal Honour done to the Place and Age, as also a great Help (if rightly considered and well apply'd) to awaken the obstinate Jews and the Vain and Dissolute Christian, who by Prophaneness, Hypocrisie and Infidelity, hath too long Blasphemed the Holy Name of Jesus.[19]

The author of the preface to the *Relation of the Miraculous Cure of Susannah Arch* took a similar approach, claiming that miracles exist "to convince an atheistical generation of men, that there is a God that acts above Power of Nature or natural causes."[20]

In his *History of the Royal Society*, the Protestant author Thomas Sprat fears that the contemporary rage for miracles will have a cheapening effect on biblical miracles: "The Enthusiastic[21] goes neer to bring down the price of the True and Primitive Miracles, by such a Vast, and such a negligent augmenting of their number."[22] Sprat believed that miracles had ceased after the establishment of Christianity. He was therefore as critical of Protestant "enthusiasts," as he called anyone who claimed divine inspiration, for their reports of miracles as he was of Catholic claims. As we have seen, Sprat's view was not accepted by most Protestants. Robert Boyle (1627–91), the so-called "father" of modern chemistry and a committed Christian, flatly denied that miracles had ceased: "I remember not, that I have hitherto met with . . . any, at least cogent, proofe that miracles were to cease with the Age of the Apostles."[23] Boyle made this comment in a letter to Henry Stubbe (1632–76), who had published an account of one of the most famous miracle cases in the seventeenth century, that of the Irish "stroker" Valentine Greatrakes (1628–83). Boyle was unhappy that Stubbe had dedicated his pamphlet about Greatrakes to him, and he wrote a long refutation of Stubbe's explanations for Greatrakes's powers. To write his refutation, Boyle became closely involved with the Greatrakes case, attending some sixty sessions in which Greatrakes "stroked" patients.[24]

Greatrakes was the talk of the coffee houses.[25] As George Rust wrote to Joseph Glanvill: "the great discourse now at the Coffee Houses, and everywhere, is about M[r] Greatrakes, the famous Irish Stroker . . . He undergoes various censures here, some take him to be a conjuror, and some an imposter, but others again adore him as an Apostle."[26] The Greatrakes case was a turning point in discussions about miracles because it represented the beginning of the enlightement debate about the plausibility of miracles, not simply their possibility. Many fellows of the Royal Society became involved in the case precisely because it touched so closely on issues that straddled the divide between the natural and the supernatural and brought up key issues involving science, religion, and magic. The dilemma facing the natural philosophers was how to frame their investigations of miracle claims, as well as any other case of the supernatural, in a way that could not be taken as a direct assault on religion.[27] As Larry Stewart puts it, "heterodoxy haunted the new Philosophy."[28] Boyle and many of his colleagues in England's Royal Society were deeply concerned by what they saw as the increasing encroachment of science on the domain of religion, and they considered this encroachment a direct cause of the growth of the skepticism and atheism threatening society. For all their interest in science and empiricism, Boyle and his colleagues were deeply religious and took it for granted that the world was full of supernatural events and forces. Not only did they attempt to document the existence of these, but they wanted to explain their causes and significance. Natural philosophers were consequently at pains to establish and affirm their orthodoxy. The very attempt, however, to winnow out false claims from true ones led to precisely what they did not want, undermining faith. The discussion about miracles involved in the Greatrakes cases encouraged the same kind of questions that witchcraft accusations and trials did about the relationship between the natural and the supernatural and reason and revelation. Such questioning was an essential component in the incremental shrinking of divine power and consequent process of secularization.

Greatrakes arrived in England in 1666. One of his first stops was at Ragley Hall, the home of Viscount Lord Conway and his wife Anne (1631–79), a highly educated woman and close friend of Henry More, the Cambridge Platonist and member of the Royal Society. Lady Conway suffered from debilitating headaches from the age of eighteen. Doctors had prescribed all kinds of noxious and toxic cures from smoking tobacco to ingesting large doses of mercury and undergoing a trepanning operation in France, which she declined at the last minute. The idea of a cure through stroking was therefore irresistible. Upon his arrival at Ragley, Greatrakes was greeted by an illustrious group of theologians and natural philosophers, among whom were the Cambridge Platonists Henry More, Ralph Cudworth, and Benjamin Whichcote, as well as George Rust, James Worthington, and Henry Stubbe. Although Greatrakes's ministrations failed to cure Anne, he racked up great success when he moved from Ragley to London. There he attracted the attention of Boyle, and crowds of curious spectators attended his dramatic performances, during which he appeared to cure cases of eczema, asthma, headaches, rheumatism, arthritis, tumors, deafness, and dropsy. Stubbe became so convinced of Greatrakes's legitimacy that he wrote a defense entitled *The Miraculous Conformist*, in which he challenged anyone to deny the evidence he had collected or cast aspersions on the credible witnesses attesting to Greatrakes's successful cures. As he says:

> I do not relate unto you the reports of interested monks and fryers concerning things done in monasteries and private cells; An infinite number of the Nobility, gentry and clergy of Warwickshire and Worcestershire, persons too understanding to be deceived, and too Honourable and worthy to deceive, will avow, that they have seen him publickly cure the lame, the blind, the deaf. The perhaps not unjustly supposed Daemonicks, and lepers; besides the Asthmas, Falling-sicknesse, convulsion-fits, Fits of the Mother,[29] Old aches and pains.[30]

For Stubbe seeing was believing: "[I] saw him put his Finger into Eares of a Man who was very thick of Hearing; and immediately he heard one when I asked him very softly severall questions." He offers a more graphic description of one cure, which provides a clear indication of the level of pain and discomfort many people routinely suffered at the time:

> I saw him launce a Wenne[31] that covered the Eye of an old man; there issued out an abundance of matter in smell and consistency, and colour, resembling a rotten-Egge; after which he crushed out the less digested matter, which resembled the braines of any creature: which being done, he stroked the place gently, and the flux of blood and pain (which was great by reason of his crushing it hard) presently ceased.[32]

Greatrakes himself wrote an account of his cures for Boyle. He was convinced that his ability to cure came from God as a sign to "convince this Age of Atheism, which (I am sorry to say it) many of our pretended wits I fear are falling into, who make it their pastime to deride Jesus and Christianity. . . ."[33] Greatrakes was a Protestant who believed in miracles. He was convinced that the ones he performed were orchestrated by God "to abate the pride of the Papists (that make miracles the undeniable Manifesto of the truth of their Church)." Thus "to make use of a Protestant to do such strange things" was God's way of helping the Protestant cause.[34] For all the Protestant protestations against miracles, during the sixteenth and seventeenth centuries there was a contest between Protestants and Catholics as to which group produced the best miracles and miraculous cures and which group should therefore win the hearts and minds of Christians.[35] Greatrakes, as we have seen, attributed his healing powers to God. This caused him to wonder why only some people were cured and not others: "Why some are cured and not all, and if this work were of God all would be cured?" Such questions, which bring up the issues of divine intervention in the world as well as divine justice, were crucial elements in the debate about miracles, magic, and witchcraft that played out in the early modern period.

Although Stubbe agreed with Greatrakes that the healings were in some cases miraculous, he had more complex explanations that combined natural and supernatural elements. God may have endowed Greatrakes with "a peculiar Temperament," but his body might also possess "particular

Ferments," and these "Ferments" might draw the "heterogeneous Ferments out of the bodies of the Diseased."[36] Drawing on the theory of fermentation devised by his Oxford tutor Thomas Willis, Stubbe offers the following convoluted "natural" explanation, an explanation that underlines the incredible difficulty facing doctors, physicians, and surgeons in their attempts to understand the complex workings of the human body in the early modern period:

> Considering that our life is but a Fermentation of the Blood, Nervous Liquor, and innate constitution of the parts of our Body, I conceive I have represented those hints and proofs which may render it imaginable that Mr. Greatrakes by his stroking may introduce an oppressed Fermentation into the Blood and Nerves, and resuscitate the oppressed nature of the parts.[37]

Henry More also suggested that Greatrakes's cures were natural. His explanation is shorter but not a whit more enlightening since he attributes the cures to a "sanative healing contagion," without specifying what that might be. He obviously discussed this idea with his patron Lord Conway, who agreed that Greatrakes's cures were not miraculous: "I am far from thinking them miracles or that his cures are at all miraculous; but I believe it is by a sanative virtue and a natural efficiency." Henry Oldenburg suggested the equally vague notion of "friction" as a cause. However, Benjamin Whichcote, whom Greatrakes had healed, rejected such natural explanations and believed Greatrakes was the instrument of God.[38] As we can see from these various theories, Greatrakes's case raised crucial questions about the boundaries between religion, magic, and science that were keenly debated at the time, making the early modern period so pivotal in the emergence of the modern world.

As mentioned earlier, Boyle was not happy that Stubbe's pamphlet about Greatrakes was addressed to him, for although he tended to think there were natural explanations for Greatrakes' cures, he found himself caught in a difficult situation. Whatever side he came down on as to whether the cures were natural or miraculous posed potential risks. In his long refutation of Stubbe he criticized him for putting Greatrakes's miracles on a par with those of Christ because that trivialized Christ's power. But he also worried that claiming the cures were natural might lead people to conclude scriptural miracles were also natural. Boyle's quandary about how to deal with Greatrakes reveals the dilemma faced by natural philosophers who offered natural explanations for what had previously been considered supernatural events. Some venturesome souls like Thomas Burnett were eager to shrink the supernatural sphere and extend the natural. As he said, "We must not fly to miracles, where man and nature are sufficient." Burnet's goal was "to see those pieces of most ancient History, which have chiefly been preserved in Scripture, conform'd anew, and by another Light, that of Nature and Philosophy."[39] This kind of thinking met with disapproval, however, from people who were apprehensive about its possible atheistical implications, precisely Boyle's concern, as we have seen. Robert Jenkins, for example, accused Burnet of having "too much Philosophy to have no Religion" and of putting "dangerous weapons into the Hands of those who have neither the one nor the other."[40]

Miracles and miraculous cures were only one area to catch the attention of natural philosophers in the early modern period. In his *Novum Organon* Francis Bacon had called for natural histories of monsters, prodigious births, and anything new or rare. London's Royal Society took up Bacon's suggestion and published in its journal, *The Philosophical Transactions*, many cases of strange phenomena or what were colloquially referred to as "wonders." These included cases of monstrous births—in one account the birth of a twenty-three-pound monster without any bones or head is described. Other "wonders" included people claiming to fast for lengthy periods and who appeared to survive without food; "old agers," whose lives were exceedingly long; and people with unusual anatomies—for example a woman in France with four breasts, a "pretty young hermaphrodite," as well as a woman pregnant for 18 years.[41] What is clear from accounts like these is that any and all examples of events with a supernatural or miraculous tinge sparked the interest of natural philosophers and stimulated the urge to investigate. In an unusual anthology, Michael Hunter presents just such a series of intriguing cases from the last two decades of the seventeenth century dealing with such disparate things as predicting the future, second sight, and Scottish fairies. What makes these documents particularly interesting is that Hunter believes Boyle initiated their collection. In his informative introductory essay Hunter claims that the subject of second sight was of special interest to English intellectuals and natural philosophers of the period because of the increasing encroachment of science on the domain of religion. He provides additional evidence that Boyle and many of his colleagues in the Royal Society were deeply concerned by the role science may have inadvertently played in promoting the skepticism and atheism they believed were increasing among their contemporaries. The debate about witchcraft was a case in point. During the seventeenth century a growing number of naturalistic and medical explanations had been given for the supposed actions of witches and spirits, and these explanations were taken by many to be a direct assault on Christianity. Hobbes, Descartes, and Spinoza were singled out as especially pernicious in this regard, for by denying the existence of spirits, they were accused of undermining the belief in God. Joseph Glanvill, a member of the Royal Society and its vociferous champion, considered a disbelief in spirits the first step in the inevitable march to atheism:

> He that thinks there is no *Witch*, believes a *Devil gratis* . . . And when men are arrived to this degree of *diffidence* and *infidelity*, we are beholden to them if they believe either *Angel* or *Spirit*, *Resurrection* of the *Body*, or *Immortality* of *Souls*. These things hang together in a *Chain* of *Connextion*, at least in these mens *Hypothesis*; and 'tis but an happy chance if he that hath lost *one link* hold another.[42]

In Hunter's view Boyle's interest in second sight was part of his more fundamental concern with promoting and affirming the truth of Christianity. Earlier in his life Boyle had subsidized the publication of an account of a poltergist in France (François Perreaud, *The Devil of Mason*, 1658), and he corresponded with Joseph Glanvill about the desirability of collecting empirically verifiable accounts of witches and spirits. (The most extensive of these appeared after Glanvill's death in *Sadducisimus Triumphatus*, 1681.) Hunter suggests that by the time Boyle interviewed the Scottish nobleman Lord Tarbat in 1678 he may have been looking for empirical evidence for the supernatural that was less sensationalist and open to charges of fraud and credulity than accounts of witchcraft.

In entitling this collection of texts "The Occult Laboratory" Hunter illustrates the significant role that antiquarian and ethnographic studies played in the religious and scientific debates of the early modern period. For in this instance the "occult laboratory" was the Scottish Highlands, which from the English point of view proved to be an exotic location for collecting and testing abnormal data. As we have seen in chapter one, European encounters with foreign peoples and places contributed to the undermining of traditional religious and scientific ideas. The documents assembled by Hunter provide an example closer to home of the way discovering new peoples, practices, and beliefs stimulated readers to reevaluate common assumptions in the light of new "factual" evidence. While those collecting these accounts hoped to use them apologetically as irrefutable proof of the supernatural, their objective was undermined by the very copiousness and questionable quality of the testimony assembled. For example, in *The Secret Commonwealth*, one of the documents in the collection, Robert Kirk presents elaborately detailed information about the nature and activities of "ELVES, FAUNES and FAIRIES." Among other things we are told that their bodies are of congealed air, that some of their books are "much like the Rosicrucian stile," that they are organized into tribes and orders, and steal human women out of child-bed to nurse their offspring. They celebrate marriages, bury their dead, speak like their human counterparts, and dress in similar plaid apparel. As Hunter points out, the harder apologists for the supernatural worked to produce their "matters of fact," the more their evidence became the target of wits and scoffers.[43]

Kirk's account also illustrates the role of ethnographic literature in introducing heterodox ideas to the European reading public. According to his report Scottish fairies do not believe in death in any conventional, Christian sense but espoused a theory of perpetual reincarnation consistent with their vitalistic view of the universe:

> For 'tis one of their Tenets, That nothing perisheth, but (as the Sun and year) everie thing goes in a Circle; Lesser or Greater, and is renewed and refreshed in it's revolutions, as 'tis another, That Every Body in the Creatione, moves (which is a sort of Life:) and that nothing moves but what has another animall moving on it, and so on, to the utmost minutest corpuscle that's capable to be a receptacle of Lyfe.[44]

One wonders if this way of thinking represents a Scottish version of the kind of religious materialism or vitalistic pantheism Carlo Ginzburg sees as characteristic of the "substratum of peasant beliefs" in the early modern period.[45] In this connection it is interesting to note that the idea of reincarnation came into vogue among English intellectuals in 1661 with the anonymous publication of *A Letter of Resolution concerning Origen and the Chief of his Opinions*. Glanvill cautiously supported the idea in his *Lux Orientalis* of the following year and Henry More followed suit. Pre-existence and reincarnation was a cornerstone of the Kabbalistic philosophy of Isaac Luria (1534–72), introduced to Europeans with the publication of the *Kabbala denudata*, or *The Kabbalah Unveiled* (1677, 1684), by the accomplished Hebrew scholar and statesman Christian Knorr von Rosenroth.[46] Kirk mentions the "preexistence of Souls, living into [*sic*] aereal vehicles."[47] Preexistence became so widely accepted that several authors felt compelled to refute the idea (Samuel Parker, *An Account of the Nature & Extent of the Divine Dominion of Goodness . . .*, 1666; E.W. *No Praexistence . . .*, 1667).

The significance of the collection of documents assembled and published by Hunter lies in their implications for understanding the evolving attitudes toward magic and the supernatural in the late seventeenth century and the effect these attitudes had in determining boundaries between the natural and the supernatural realms. Underlying the interest in second sight were the same issues that surface in the debate over magic, miracles, and witchcraft concerning the authority and credibility of the Christian revelation, the role of God and spirits in the physical universe, and the epistemological problem of what constitutes legitimate scientific knowledge. The investigation of phenomena such as miracles, miraculous cures, magic, and second sight was therefore not an anomalous aspect of the period of the Scientific Revolution but an integral part of it. To arrive at the modern definition of what constitutes a scientific "fact" or "theory" required new concepts of what constituted valid scientific evidence together with reasonable and convincing explanations. In this regard it is rewarding to remember the confusing mix of natural and supernatural proofs and explanations given to explain Greatrakes's cures and to consider the various theories accounting for the phenomenon of second sight. The fact that second sight is mentioned in the Bible and confirmed by numerous and reliable ancient and contemporary sources is taken as adequate proof by Kirk and many others, including Samuel Pepys.[48] However, such second-hand reports were no longer sufficient to convince Lord Tarbat, who preferred his own eye-witness observations: "I will choose rather to put myself than my friends on the hazard of being laughed at for Incredible Relationes."[49] Yet, Lord Tarbat included many second-hand accounts of second sight in his report and seemed convinced by them, which shows just how vexed the question of what counted as reliable scientific evidence continued to be throughout the early modern period.

A wide variety of supernatural and natural explanations are suggested for how and why people have second sight: it can be obtained by means of a magical rite, in which "hair (which bound a Corps to the Bier)" is run around a person's body and he looks downward, "as did Elijah 1 King 18.42," and "back thorow his legs until he see a funerall advance . . . or back thorow a hole where was a knot of fir."[50] Alternatively, second sight can be obtained through contact with someone already possessing the skill, as a gift of good or bad spirits, or simply by paying someone to teach the skill "for a pound or 2 of tobacco."[51] In other instances it is said to be hereditary and a special prerogative of "Seventh-sons." In this case it may result from "miraculous operations" or "some secret virtue in the womb of the parent," but it may also proceed naturally "onlie from the Sanative Balsome of their healthful constitutions." Although he was at a loss to explain "the cause of so extraordinary a phaenomenon," Lord Tarbat vaguely noted that it might be due to "a quality in the eyes of some people in those parts, concurring with a quality in the air also."[52] In line with these more naturalistic explanations is the idea that for some people second sight may simply be "acquired as an artificiall improvement of their natural sight . . . ; Resembling in their own kind, the usuall artificiall helps of Optic Glasses (as prospectives, Telescopes, and Microscopes). . . . "[53] However unconvincing these explanations may be, evaluating the varied kinds of testimony adduced to prove the reality of second sight as well as the explanations for it were essential steps in the development of modern science and scientific methodology.

St. Paul's contention that "we see through a glass darkly" may well have applied to his own time, but it was never more evident than in the sixteenth and seventeenth centuries when theologians, demonologists, inquisitors, and natural philosophers struggled with the problem of distinguishing what was real and knowable from what was imaginary and false. The awful thought that seeing was not believing and, conversely, that beliefs could not literally be "seen" led to a crisis of skepticism that played itself out in inquisitorial courts, the book-lined studies of theologians, the laboratories of natural philosophers, and over the naked bodies of suspect witches. To this last subject we now turn.

CHAPTER 4

WITCHES AND WITCH HUNTING

By the time William Hogarth published his engraving "Credulity, Superstition and Fanaticism" in 1762, the belief in witches had all but disappeared. Joseph Glanvill's once famous book *Saducisimus Triumphatus: or Full and Plain evidence concerning Witches and Apparitions* (1668), published with Henry More's ample annotations three times in the seventeenth century and twice more in the eighteenth, was an object of curiosity, a relic of those past popular delusions ridiculed in the age of Enlightenment. Once credited with putting "the belief in apparitions and witchcraft on an unshakable basis of science and philosophy,"[1] Glanvill's book appears in Hogarth's print as the ultimate source of that credulity, superstition, and fanaticism delineated by the artist so carefully and critically. Placed in the bottom right-hand corner, Glanvill's work provides a platform, first, for Wesley's sermons and, then, for a human heart in which a thermometer has been inserted with degrees of heat registered by passions and mental disorders. The scale starts with suicide, madness, and despair and ends in lust, ecstasy, convulsive fits, and raving. Superstition and credulity, represented by Glanvill and Wesley, provide the foundation for the varying scenes of insanity depicted in the print itself. Many of the portraits come right out of Glanvill. The shoe-black in the foreground, for example, vomits nails and pins, a sign she is bewitched. She holds a bottle of urine, in which she has attempted to confine the evil spirit possessing her, but the cork has popped out, allowing the spirit to escape. Just such a scene appears in a section of Glanvill's book added by Henry More.[2]Another scene in the left foreground features a more recent case reported in 1726, involving a twenty-five-year-old maidservant by the name of Mary Toft. Mary claimed to have given birth to sixteen rabbits. By all accounts she

William Hogarth, *Credulity, Superstition and Fanaticism* (1762). (Courtesy of the Trustees of the British Museum.)

was obsessed with rabbits. While pregnant she had been observed admiring rabbits in the local market and chasing rabbits from her garden, and she had developed an unnatural craving for roast rabbit. Her obsession gave support to the prevailing theory that fetuses could be affected by strong impressions made on the mother. In this case, strong rabbit impressions transformed Mary's growing fetus into fuzzy bunnies.

Hogarth had a field day with Mary Toft and featured her story in several prints to satirize the religious and scientific superstitions of the day. His engraving "Cunicularii" is not only a satirical take on Mary's claims, but a burlesque of the Virgin Birth. Playing on the similarity of the word for rabbit, "cuniculus," and the word for vulva, "cunnus," Hogarth has one of the three doctors witnessing the scene exclaim, "A great birth." His engraving draws on traditional scenes of the three wise men bearing gifts to the Christ child with Mary Toft standing in for the Virgin and the rabbits for Jesus.[3] Hogarth's target in this satire was both religious and scientific "enthusiasm," a topic of great concern at the time when it was hard, if not impossible in many cases, to distinguish fact from fiction. The story of Mary Toft caused a sensation as pamphlets and tracts for and against the truth of her extraordinary experience proliferated. Evaluating the case, Lisa Forman Cody points out that those who believed in the rabbit births used the same standards of evidence as those who rejected it as a fraud. How then was one to know what or whom to believe?

The purpose of this chapter is to investigate the mental world satirized in Hogarth's prints. Whatever Hogarth may have thought, Joseph Glanvill (1636–80) was not a fanatic, but a highly educated and intelligent proponent of the new science and a member of the Royal Society. Henry More (1614–87), one of the most famous of the so-called Cambridge Platonists, was also a member of the Royal Society, and both men were advocates of experimental philosophy. At the time they lived and wrote there was nothing unusual about the synthesis of science and demonology to which both men subscribed. To think there was ignores the fact that the height of the witch hunts occurred between 1570 and 1680, a period that overlaps with what is generally known as the Scientific Revolution. During this period some 35,000 individuals, the overwhelming majority women, were stripped naked and "pricked" by special court officials with special instruments to see if their bodies possessed insensitive spots known as "devil marks." The Jesuit Martin Del Rio claimed that devil marks might look like a spider, a cat, the footprint of a toad or hare, or simply a mole or flea bite. The places to look for devil marks on men were under the eyelids, the lips, armpits, and rectum. In women devil marks were usually found on the breasts or genitals.[4] If such marks were discovered, the suspects were tortured (except in England where torture was outlawed in cases of witchcraft, although occasionally practiced) until they confessed they were the devil's disciples, at which point they were burned at the stake or hanged.[5] In 1602 the French demonologist Henri Boguet (c. 1550–1619) claimed that an army of 1,800,000 witches threatened Europe's, an army of a size never seen throughout Europe long history of warfare.[6] Boguet was a lesser light among a galaxy of geniuses who believed in witchcraft and actively supported witch hunts. For the most part witch hunters and demonologists were scholars and rationalists, cultivated, erudite, and eminently respected and respectable men. Martin del Rio (1551–1608), for example, who was so knowledgeable about devil's marks, was described by a contemporary as "the miracle of our age." While still in his teens, he devised a special tricycle with a desk top that enabled him to scoot between book shelves in the great libraries of Europe. With this labor-saving device he published an edition of Seneca at the age of nineteen, citing 1,100 authors.[7] Jean Bodin (1530–96) was another implacable witch hunter of profound erudition. He was enlightened enough to suggest that increased almsgiving would eradicate witch beliefs; nevertheless, he believed that customary laws of evidence should be suspended in witchcraft cases. In his opinion, circumstantial evidence of a supernumerary teat or nipple (9% of the world's population has these), a wart, an ugly face, and bad body odor (the percentage of the population with any or all of these was staggering in an age of rampant small pox, little concern or knowledge of hygiene, no deodorants, and no tooth brushes or tooth paste, not to mention teeth!) were sufficient to consign suspects of witchcraft to the torture chamber.[8]

During the past three decades there has been an increasing recognition of the historical importance of witch hunts in the emergence of the modern world. Scholars have come to realize that understanding such pivotal events as the Protestant Reformation, the Catholic Counter Reformation, the Scientific Revolution, and the Enlightenment will remain incomplete until historians have fully investigated the role that witch hunts played in shaping modern forms of social and cultural identity.[9] Witch hunts were an important element in the process of centralizing both church and state. They stemmed in large part from the "confessionalization" of Europe that did so much to encourage religious fervor and the commitment to stamp out heresy on the part of Protestants and Catholics alike. Witch hunts reflected the confrontation between learned and popular culture accompanying this "confessionalization," a confrontation that generated stricter ideals of moral discipline and new standards of social and cultural behavior.[10] Witch hunts were also of great importance in the history of women, for they represented an unparalleled form of misogyny, in which women were demonized as a group and perceived as a diabolical fifth column threatening the very existence of Western civilization.[11] Witch hunts produced an essentialist rhetoric of gender identity that polarized the sexes as never before. They laid the foundation for the kind of gender "complementarity" that Thomas Laqueur and Londa Schiebinger claim only emerges in the second half of the eighteenth century.[12] While the idea of gender complementarity certainly did come to the fore during this later period—for different reasons and in different ways—it was fully in place at a much earlier date, although it might be more accurate to replace the term "complementarity" with "contrariety," following Stuart Clark.[13] Whether women were considered the complements of men or their contraries, the fact that they were thought of as being fundamentally different had a tremendous impact in legitimizing witch hunting and, more importantly, the conviction that women were morally and intellectually inferior to men.

Witch hunts were also an important aspect of the emergence of modern science in two major respects. First, they call into question the idea that early modern "natural philosophy" exhibited a more positive attitude toward "female" nature (and by implication toward women) than modern science. The idea that modern science emerged when women were excluded and science itself was "masculinized" ignores the fact that this exclusionary process began much earlier and was not a new feature of the eighteenth century. Early modern natural philosophers were capable of incorporating into their scientific theories the same kind of misogyny characteristic of the most committed demonologists and witch theorists, who were themselves students of natural philosophy.[14] Second, the debate over witchcraft concerned questions about the authority and credibility of the Christian revelation, the role of God and spirits in the physical universe, and the epistemological problem of what constitutes valid scientific knowledge. The modern definition of what constitutes a scientific "fact" or "theory" was hammered out during the early modern period, as were new concepts about what constituted valid scientific evidence and convincing scientific explanations. That these debates entailed the execution of some 29,000 women charged with witchcraft places women and their bodies at the center of a major cultural transformation from the pre-modern to the modern world.

As we have seen in previous chapters, one of the most obvious features of the Reformation and Counter-Reformation was a new and unprecedented concern with order and orthodoxy in the face of confusion and uncertainty. The collapse of religious consensus was paralleled by the breakdown of traditional intellectual and scientific systems. The fascination with everything "unnatural" and "abnormal" was indicative of the profound anxiety caused by the destruction of existing categories and the attempt to impose, or create, new categories without ambiguities. This is especially noticeable in the area of sexuality and gender identity.

The fact that the most intense period of witch hunting occurred during the sixteenth and seventeenth centuries illustrates the centrality of gender issues at the time. Unlike Stuart Clark, who has argued that it is "a question mal posé to ask why women were the main objects of witch persecutions,"[15] one might suggest that this is and always has been the crucial question. Clark dismisses it as tautological on the grounds that the polarity between the genders was so firmly in place during the period of the witch hunts, that witches were by definition women, and demonologists had "no choice" but to define them as such.[16] It is certainly true that a polarized view of the sexes was a part of Europe's inheritance from classical sources, particularly Aristotle and Hippocrates; but it was only in the early modern period that this polarity became firmly entrenched in learned discourse. Clark seems to recognize this. As he says, "In the case of the sixteenth and seventeenth centuries, one is struck forcibly by the profusion—almost a promiscuity—of indigenous styles of oppositional thought and writing, and by a delight in listing and exploring the binary aspects of experience."[17] Clark makes the further point that the various explanations suggested for why women were targeted as witches (the social marginality of poor, old, single women; the anomalous position of women who inherited male property; the increasing number of unmarried women) "have not shown . . . [and] cannot show . . . why the accusations should have concerned witchcraft rather than some other crime."[18] But I would argue that Clark has the argument backwards. Rather than accepting the identification of woman with witch as unproblematic, what needs to be shown is, first, why witchcraft emerged as such a threatening possibility in the early modern period and, then, how the emerging definition of a bad woman dovetailed with the emerging image of a new kind of witch specially tailored in response to early modern problems and concerns.

Historians have described the pessimism characterizing the Reformation and Counter-Reformation. The bitter religious warfare endemic to the period combined with severe famines, peasant unrest, and the economic changes connected to the emergence of proto-capitalism convinced many people they were experiencing the tumultuous times preceding the apocalypse. In this atmosphere of fear of the devil and his minions, witch panics proliferated and the tendency to demonize one's enemies became common. Before the late fifteenth century witches were not identified predominantly as women. They could be male or female, and they were often well-born, if not noble.[19] By the late sixteenth century, however, between 71–92 percent of those accused in most places were women. Thus during the time that witch hunts were at their peak, the most conspicuous attribute of witches was their gender. In her study of American witchcraft,

Carol Karlsen found that men who confessed to the crime of witchcraft were often rebuked as liars, whereas women were taken at their word and executed.[20] Erik Midelfort noted a similar tendency of males to believe and convict females.[21] Another salient characteristic of the witch was her age. Women over forty were more likely to be accused of witchcraft and, after being accused, to be executed.[22] In short, the stereotype of the witch as an old, querulous crone is accurate up to a point. However, one must add that the most likely candidates for witchcraft accusations were those women who did not fit the masculine stereotype of the good woman as an obedient, silent, and submissive wife and mother, dependent on male kin. The majority of witches were past child-bearing age and a good percentage were unmarried, widowed, or living alone.[23] Among the younger witch suspects a significant number were charged with sexual crimes—fornication, adultery, abortion, or infanticide—or had given birth to illegitimate children. Witchcraft was the most important capital crime for women in early modern history. Between 1480 and 1700 more women were executed for this crime than for all other crimes put together.

The preponderance of women among those accused and executed for witchcraft makes it all the more astonishing that until relatively recently historians have denied that witch hunts had anything to do with gender. As one scholar of English witchcraft concluded, "There is no evidence that hostility between the sexes lay behind their prosecutions."[24] The general consensus was that witches pretty much caused their own problems because they were "mad," "weird," "quarrelsome," sexually inadequate, or deviant.[25] Only in the past few decades have historians concluded that misogyny is central to understanding witch hunts, and this insight occurred primarily because feminist scholars made gender a fundamental category of historical analysis.[26] Historians have increasingly come to realize that people were encouraged to believe in the existence of diabolical female witches because of what might be described as the gradual demonization of women, a process that began in earnest in the twelfth century and escalated dramatically during the period of witch hunting.

The sixteenth century has been described as one of the most bitterly misogynist periods in Western history.[27] By rejecting celibacy as a legitimate alternative to marriage, by confining sex within marriage more strictly than ever before, and by suggesting that sexual satisfaction was a legitimate concern for wives as well as husbands, Protestantism increased male sexual anxieties. Luther's remark about how bizarre it was for him to wake up and see his wife's braids on the pillow beside him is indicative of the radical reassessment Protestant theologians had to make about marriage and gender relations. In their initial stages both Lutheranism and Calvinism tried to eliminate the double standard in sexual mores by making men adhere to the same ideal of chastity before marriage and fidelity during marriage prescribed for women.[28] Such an ideal made illicit sexual activity seem even worse, and with the abolition of the confessional male guilt over sexual transgressions escalated. Guilt leads to projection, to the transfer of responsibility from the perpetuator to the other party and to women in general. The polemical debate among Protestants about the pros and cons of marriage reveals considerable projection, which expressed itself in graphic descriptions of the repellant, untrustworthy, and dangerous nature of women.[29] In this respect the witch was a cautionary figure, refurbished and in many respects reinvented—or perhaps one should say invented—in the early modern period to keep women in their place. For witches were precisely those women who refused their newly prescribed role as submissive wives and mothers. The witch was the antithesis of the good wife. She was verbally abusive, when she should be silent, promiscuous—and homosexually promiscuous at that—when she should be chaste, domineering when she should be obedient, and out and about, when she should be at home. In short, witches were women who rejected the private world of female domesticity for the public world of men. They were women who rebelled, and on the basis of 1 Samuel 15:23 ("for rebellion is as the sin of witchcraft") rebellion was routinely equated with witchcraft and rebellious wives with witches. In 1692 William Good told one of the Salem judges that "he was afraid that [his wife Sarah] either was a witch or would be one very quickly" because of "her bad carriage to him."[30]

The figure of the witch was held up to women in sermons, devil books, and plays as a deterrent, but the antithesis between the good wife and evil witch masked the very real male fear that deep down all women were potential witches. They were so because of their subservient position. Thomas Cooper makes this point in his treatise *The Mystery of Witchcraft*. Women are, he claims, "usually more ambitious and desirous of Soveraignety the rather because they are bound to subiection."[31] Cooper recognized what Freud did not, that because women are disenfranchised in a patriarchal system, they are bound to work against it. This did not, however, make him sympathetic to their plight, just more fearful of them. Even submissive and obedient wives found themselves in a quandary, for the behavior demanded of them by patriarchal society encouraged traits that belong more properly to a witch than to a good wife. Wives were expected to "cajole," "charm," and "entice" their husbands away from evil thoughts and evil deeds. As Luther says, wives "should deport themselves in such a way in the matter of gestures and conduct that they entice [reytzen] their husbands to believe."[32] While the ends of a wife might be entirely different from those of a witch, their means are uncannily alike.

The heightened fear of women and deviant female sexuality are central motifs in the iconography of the Witches' Sabbath. This did not exist in the Middle Ages; it was an invention of early modern demonologists and represented an increased fear and fascination with the whole gamut of female sexuality on the part of Catholics and Protestants alike.[33] The obsession with female sexuality accounts for the crackdown on prostitution and the outlawing of brothels in the late medieval and early modern periods. Procuring, concubinage, and adultery, together with dancing, dress, and ornament, were all attacked by preachers and moralists and regulated by governments.[34] But the image that inspired the greatest fear of female sexual deviancy was the witch as she met her cohorts in wild, mountainous lairs, where they indulged in obscene sexual activity with the devil, danced lasciviously, and feasted on noxious food and drink brewed from loathsome animals, reptiles, and insects with the addition of infant body parts and male penises.

The new group identity and group activities of witches corroborated the deepest fears of men. For women played an active role in the early years of both the Protestant and Catholic Reformations, and while their participation was initially encouraged, it was discouraged once Protestantism was established and Catholic renewal in place. At that point males were eager to reassert their authority, but women were not as eager to accept it.[35] Amy Leonard sums up the situation: "although this was a period of a flowering of female spirituality, it was also a time of increasing panic about women's freedom and activities outside of the cloister or marriage."[36] So great was this panic that in a final decree (1563) the Council of Trent stipulated that monastic enclosure was to be the rule for all religious women. The reaction of the Protestant male hierarchy to women who taught and preached was equally severe. In the view of many Protestant leaders these women were no better than witches because they rebelled against male authority. The intemperate language with which the Puritan minister John Cotton denounced Anne Hutchison, his one time close friend, for daring to hold religious meetings reveals that he considered these meetings as bad as any witches' Sabbath:

> You cannot evade the Argument . . . that fillthie Sinne of the Communitie of Woemen; and all promiscuous and filthie comings together of men and Woemen without distinction or Relation of Marriage, will necessarily follow. . . . Though I have not heard, nayther do I thinke you have been unfaythfull to your Husband in his Marriage Covenant, yet that will follow upon it.[37]

The Protestant magistrate Hugh Peters was more troubled by Anne's masculine behavior than by her antinomian opinions because such behavior undermined the natural superiority of the husband in the family unit. As Peters said, "You have stept out of your place. You have rather been a husband than a wife; and a preacher than a hearer; and a magistrate than a subject. . . . "[38] Catholic authorities were no less suspicious of active, independent women. Although they wanted to inspire women spiritually, they tried to restrain them as well and keep them firmly under male control and supervision. Teresa of Avila (d. 1582) is a case in point. The mystical writings that have made her justly famous were minutely scrutinized by church officials. Her life in the world rather than the cloister caused so much consternation that one papal legate described her as a

> restless wanderer, disobedient, and stubborn *femina* who, under the title of devotion, invented bad doctrines, moving outside the cloister against the rule of the Council of Trent and her prelates; teaching as a master against St. Paul's orders that women should not teach.[39]

Although she was eventually canonized, she was called upon to defend herself before the Spanish Inquisition.

Catholic wives, like Protestant ones, were also subject to supervision and control of their husbands. Addressing "her girls" at the Saint-Cyr, Mme de Maintenon had extremely direct and telling information to impart to them on the subject of marriage. There was nothing romantic in the vision she presented: women could not expect to have any life of their own once married:

> You will have, Mademoiselle, your husband to look after, and then you will have a master. . . . You may displease him; he may displease

> you. It is virtually impossible that your tastes will be similar. He may be of a mind to ruin you; he may be greedy and deny you everything. It would be tedious if I told you what marriage was like.[40]

If this bit of wisdom came from the wife of Louis XIV, the Sun King, what could the wives or more ordinary male mortals expect? To those who complained about school discipline, Mme de Maintenon had the last word, warning that it was nothing in comparison to what lay in store for them:

> To girls who believed that no discipline could be more severe than that of their school, Mme de Maintenon affirmed that the rule of family life was far harsher. A married woman could count on no rest, no secrecy, no privacy of her own, for her services were continually and urgently in demand by husband, children, dependents, and servants. She had no more private life than "a sergeant traveling from skirmish to skirmish" in the thick of battle—and battles were relatively infrequent, whereas the married woman was constantly under fire.[41]

The tough (and by modern standards inhumane) regimen described by Mme de Maintenon for wives underlines the significance of the witch hunts in the history of women. The changing iconography of the witch from the medieval to the early modern period offers vivid proof of the heightened fear disruptive women inspired and helps to explain the reason for the draconian measures taken, or at least suggested, to alleviate the threat they ostensibly posed. Late fifteenth-century illustrations of witches show them as both male and female and fully clothed as they engaged in various nefarious activities such as casting spells, harming animals and humans, and causing storms. But around the beginning of the sixteenth century witchcraft illustrations focus on the bodies and sexuality of female witches.[42] Durer's witch is a case in point. Not only does she ride backwards, an image of disorder, especially sexual disorder, but the distaff that should identify her as a submissive female is suggestive of a phallus. This together with the fact that she clutches the horn of the goat on which she rides implies that she has appropriated and controls male sexual power. Dürer's witch is a cautionary figure of the dangers presented by disorderly women.[43] During the sixteenth century the theme of female sexual power and the reversal of gender roles was extraordinarily widespread. Illustrations of Solomon worshipping idols under the influence of his foreign wives, Samson having his hair cut by

Albrecht Dürer, Witch riding backwards on a goat (c. 1500). (Ernst and Johanna Lehner, *Picture Book of Devils, Demons and Witchcraft*. New York: Dover Publications, Inc., 1971, 65.)

Delilah (one doesn't have to be a Freudian to see this as symbolic castration), David and Bathsheba, Aristotle and Phyllis, and Hercules dressed in Omphale's clothes and carrying a distaff were widely disseminated images of "the woman on top."[44] The connection between female sexuality and sin was so pronounced that visual representations of the Fall lay the blame squarely on Eve's sexuality.[45] Images juxtaposing women, sex, sin, and death were commonplace. The figure of the witch embodied all these motifs, as one can see in the stunning series of erotic images of naked witches by Hans Baldung. A comparison of Baldung's depictions of witches with those of Ulrich Molitor, whose work predates Baldung's by some twenty years, is instructive. Molitor's witches are fully dressed and not the least erotic. They do not roast suggestive

A female witch laming a man. (From Ulrich Molitor, *De lamiis et phitonicis mulieribus*. Cologne, 1489). From Ernst and Johanna Lehner, *Picture Book of Devils, Demons and Witchcraft*. (New York: Dover Publications, Inc., 1971, 60.)

looking "sausages" as Baldung's witches do, a reference to the new view of witches as relentless castrators.[46]

The seriousness of all this material does not call into question the fact that deviant female sexuality had always been a subject of great mirth as well as horror and concern. One has only to read medieval fabliaux or Renaissance texts devoted to the "querelle des femmes" to realize that misogyny transcended religious and national boundaries. But the quantity and viciousness of misogynist invective in the early modern period was unparalleled. Moreover, the epicenter of this misogyny was Germany, where Lutheranism began, the witch panics were most intense, books about the devil most popular, and executions for witchcraft most numerous. Probably more witches were executed within the boundaries of present-day Germany than in the rest of Europe put together.[47] Sixteenth- and seventeenth-century German broadsheets are filled with the marital woes of model husbands. Disorderly wives beat and trick their husbands, drink excessively, feast extravagantly, ignore housework, take lovers, and consult witches. The image of the woman on top and wearing men's breeches was as common as it was cautionary; and it was a universal maxim that women were sexually rapacious. One broad sheet offers a

Hans Baldung, Witches concocting an ointment to be used for flying to the Sabbath (1514). (Ernst and Johanna Lehner, *Picture Book of Devils, Demons and Witchcraft*. New York: Dover Publications, Inc., 1971, 63.)

sardonic "summary of how every woman with a wretched, dissolute husband shall lick him with sticks until his ass-hole is roaring."[48]

Husband abuse and husband beating were clearly hot topics, which belies the fact that it was far more likely to be the other way around. The latitude given to husbands to discipline their wives in the early modern period is astonishing by modern Western legal standards and the level of brutality allowed and even encouraged breathtaking. Sixteenth-century broadsheets contain fulsome advice about how husband should deal with obstreperous wives. In one, "Rust's well-tested recipe to cure the evil disease of disobedient wives," the husband solves his marital problems by beating his wife to death. His solution is not viewed as excessive, but eminently just. In the last scene he is in a tavern celebrating his release from an impossible marriage with a shrewish, disobedient wife as her funeral cortege passes by the open door.[49] Beating was so routinely recommended to husbands as the solution for their marital woes that the allegorical figure of Dr. Kolbmann, or Dr. Clubman, emerges in popular literature. The following ditty advocating wife-beating cut across national boundaries and became something of an international conceit:

> Do you have an evil wife on Sunday?
> Go into the woods on Monday.
> Cut a switch on Tuesday.
> Beat her with it on Wednesday.
> She will be ill on Thursday.
> Let her lie there on Friday.
> She will then die on Saturday.
> She will be buried on Sunday.
> And in this way you will have a good Monday.[50]

This light-hearted bit of male folk wisdom fit in well with the sixteenth-century commonplace that a married man has only two days of happiness, the day he married and the day he buries his wife.

It might be objected that all of this is simply meant in jest. Unfortunately it was not as is evident from the laws condoning wife-beating promulgated in the Middle Ages but still on the books during the period of the witch hunts. The common law of Beauvais, for example, allowed a man to beat his wife "when she refused her husband anything." A law of Bergerac permitted a husband to draw blood as long as he did it with "bono zelo" (good zeal). Customary law in Bordeaux went so far as to exonerate a husband who killed his wife in a fit of rage, but only if he confessed under oath that he was repentant.[51] English law on wife beating was more subtle. It was legal for a husband to beat his wife unconscious, but not to the point that her inert body farted, a sign she was in shock and possibly dying.[52] Wife beating was so prevalent in sixteenth-century London that civic regulations forbade it after nine in the evening because the noise disturbed neighbors.[53] Protestant authorities, like Catholic ones, took it for granted that husbands had the duty to chastise disobedient wives. Even in cases of severe abuse, Protestant authorities were reluctant to sanction a wife's request for separation or divorce.[54]

Given the supposedly parlous state of husbands it is not surprising that the figures of Bigorne and Chichvache enjoyed renewed popularity in the sixteenth century. Chichevache first appears in an anonymous French poem early in the fourteenth century and is mentioned by Chaucer as well.[55] Bigorne appeared about the same time. Both summed up marriage from the male point of view. Bigorne, colloquially known as "Fillgut," waxed monstrously fat on a diet of obedient husbands, while Chichevache, or "Pinchbelly," starved on a diet of virtuous wives. In the sixteenth century Chichevache became an eater of virgins, which explains his even greater emaciation. What had been a medieval satire on shrewish women became a Reformation satire on all women, a shift of emphasis reflecting the increased concern with the uncontrollable nature of females and female sexuality.

As this review of attitudes toward women makes clear, the notion of wives as devious and dangerous was not an early modern invention but a commonplace throughout Europe during the Middle Ages. But in the medieval period, this literature was counterbalanced by courtly literature with its idealization of women and love and the recognition of the transforming effects of both on male conduct. No such literary counterbalance exists in the early modern period. Sex is dangerous, and the dangerous sex is female. Female sexual deviancy became an obsession of male authorities and men at large, and this explains why witch theorists expended so much time and effort interrogating witch suspects about the precise nature of their sexual encounters with the devil and demonic spirits, why prostitutes were demonized, prosecuted, and expelled from towns and cities that had previously accepted them as a necessary outlets for male libidos, why the number of women charged with sexual crimes increased exponentially, and why magistrates wanted ever-more salacious details about the sex lives of those women they deemed deviant.[56] The figure of the sexually voracious women was endlessly elaborated and became an effective defense for males charged with sexual misconduct. Roper discusses a case in which a male was sentenced to four weeks in prison, while his female partner in fornication was given the much harsher punishment of exile simply because he claimed that "she had sprung on him like a billygoat."[57] So conscious were males of the unbridled nature of female sexuality that even sewing-bees were looked upon with suspicion,[58] and women were warned against the practice of solitary reading for fear it would lead to masturbation.[59]

According to Hans Peter Broedel the kind of emasculating witch envisioned by Dürer was largely invented by the Dominican inquisitors Henrich Kramer (Henricus Istitoris)(1430–1505) and Jacob Sprenger (1436/1438–94), authors of the notorious *Malleus Maleficarum* (*The Hammer of Witches*) of 1486, a handbook for inquisitors that set down everything one needed to know about the evil activities of witches and the way to prosecute them.[60] There was no ambiguity in the *Malleus* about the fact that witches were women. The very title makes this clear because *maleficarum* is a feminine noun in Latin. The *Malleus* has been described as "scholastic pornography," although one of the more recent books on witchcraft denies this, as we shall see.[61] The authors clearly dreaded and hated women. Witches were not only lying, cheating, credulous, vain, ambitious, untrustworthy, and lustful; they were also relentless castrators, as the following passage reveals. If this had not been written in deadly earnest and had deadly effects, it would be worthy of Monty Python:

> So what are we to think about those witches who shut up penises in what are sometimes prolific numbers, twenty or thirty at a single time, in a bird's nest or some kind of box, where they move about in order to eat oats and fodder, as though they were alive—something which many people have seen and is reported by common gossip? It must be said that all this is done by devilish activity and by hallucination, and this is how the senses of those who see [such things] are deluded in the ways I have described earlier. For example, a man reported that when he had lost his penis, he approached a witch in order to get himself made whole again. She instructed the sick man to climb a particular tree and, in a kindly way, told him he could take any one he wanted from a nest in which there was a very large number of penises. When he tried to take a big one, the witch said, 'Please don't take that one,' and added that it belonged to one of the parish priests.[62]

What is to be thought of men who could credit such a preposterous story? Where did Kramer and Sprenger get such incredible information? How did they and the fellow inquisitors and judges influenced by their book come to have such clear and detailed knowledge about the sexual activities of witches? And why was sex so central to their concerns rather than the harmful actions of witches, which were of far more interest to the common people who claimed to suffer from them? For answers to these essential questions, we can turn to Walter Stephens's insightful book *Demon Lovers*.

Stephens contends that witchcraft theorists were neither credulous fools nor prurient misogynists, but tormented skeptics trying to resolve the conflicts in Christian doctrine about the benevolence of God, the existence of spirits and souls, and the efficacy of the sacraments. Witchcraft theory was consequently a kind of "theological damage control" in the face of increasing skepticism.[63] Stephens sees 1400—the same date an increasing concern with demonic magic emerged—as pivotal, marking the point when significant numbers of educated Christians began to believe that human beings, especially women, interacted with demons in intensely physical ways, the most pronounced of which was through sexual intercourse. The Black Death, the Great Schism, the discovery and dissemination of new texts, printing, trade, travel, and the discovery of the new world all undermined established truths and called into question the idea of divine providence and God's omniscience and benevolence. Misfortune, uncertainty, and insecurity called for a new theodicy, and Stephens argues that witch theorists supplied this. A consistent theme runs through all their writings: the terrible fear that God, spirits, heaven, and hell did not exist. They wrote to assuage their deepest doubts, and these doubts could be kept at bay only by proving that spirits were real and interacted on a physical level with human beings.

Stephens claims that the reason why witch theorists were so interested in proving the reality of demonic sex had little to do with an obsession with sex or misogyny but was a consequence of their attempt to provide irrefutable evidence that demons did interact physically with humans. What, after all, could be better proof of physical intimacy than intercourse?

> Literate interest in copulation with demons was hardly driven by prurience, misogyny, or puritanical fervor. Literate men craved demonstrations that even sexual intercourse, the most intimate sort of bodily contact, was possible with demons. Copulation offered valuable perspectives on the life of demons, their corporeality, and the possibility of acting meaningfully with them.[64]

Witch theorists were "metaphysical voyeurs"; their deepest desire was not to look through bedroom keyholes but through those barriers that separated the physical world of human beings from the spiritual life to come.[65] While Stephens admits that witch theorists were misogynists, he claims they did not exploit theology to demonize women but utilized the prevailing misogyny to reinforce a demonology that supported the religion they seriously doubted. This is a valid point, but only up to a point. Stephens's valiant attempt to apply this logic to Heinrich Kramer and to exculpate him from the charge of writing pornography is ultimately unconvincing:

> Of course the *Malleus* was misogynistic, but what for Kramer was the use of misogyny? To read his treatment of demonic copulation as a tirade against women's sexual powers is to miss his point entirely. If anything, his tirade is for women's sexuality. The issue was not keeping women in their place or controlling their sexuality. Henrich Kramer did not fear that women were associating with demons: he hoped that they were. His whole theology depended on women's sexual transgressions, and it would have collapsed if he had ever had to admit that women's behavior conformed to the patriarchal ideal of chastity and submissiveness.[66]

It is hard to believe that the long and tortured history of Christian ambivalence to sex and procreation and the misogyny this generated were not important contributing factors to witch theorists' emphasis on the perverse sexuality of female witches. Stephens's explanation that demonic sex replaced *maleficum* (an evil deed or harmful magic) as the defining characteristic of witchcraft because sexual intercourse offered better proof of demonic and human interaction—"Kramer's interest is not prurience but provability"—is problematic.[67] Surely, an emphasis on witches and demons making demonic pacts could have provided the same level of "proof" that physical interaction was possible. The fact that witch theorists were so interested in the salacious details of the sexual encounters between witches and demons, so detailed in their descriptions of the size and shape of demonic sex organs (enormous), so curious as to whether women preferred sex with demons or humans, so convinced of the painful and unpleasant nature of demonic intercourse, and so insistent that even demons refrained from "crimes against nature" (i.e., sodomy or even intercourse in anything but the missionary position) indicates a fixation on sexuality and gender profoundly colored by Christian attitudes. It also suggests that there was an increase in the level of misogyny in the late medieval and early modern period that made it possible for female witches to replace male necromancers, heretics, and Jews as the Church's most dreaded enemy. It is at this point that it is useful to return to Broedel's analysis of the *Malleus*.

Although Broedel is prepared to admit that Kramer's and Sprenger's view of women was deeply misogynist, he agrees with Stephens to a certain extent by arguing that their identification of witches as female "is quite probably descriptive rather than prescriptive" inasmuch as it represents popular opinion.[68] In Broedel's view, the genius—or perhaps it would be better to say the evil genius—of Kramer and Sprenger's work is that it reconciled learned and popular views of witchcraft. As scholars have long realized, there was a gulf between what educated theologians thought about witches and the views of less well-educated or uneducated laypeople. Before the publication of the *Malleus* witch theorists and demonologists tended to view witches as heretics, allies of the devil bent on attacking the Church and enticing its members into sin and damnation. Ordinary people took a very different view. For them, witches were primarily women with the power to inflict harm (through an act of *maleficium*); the relation of these women with the devil was not an issue. What ordinary people wanted to know is why bad things happened to them, and the witch supplied an ideal scapegoat in a society that looked for personal explanations for misfortune. In their role as Inquisitors, Kramer and Sprenger were continually confronted by this popular view of witches and witchcraft. But because it contradicted accepted authorities, who considered witches delusional and denied them any power, they felt obliged to construct a new model of witchcraft that reconciled learned and popular beliefs.

Broedel argues that their new model was predicated on Aquinas's view of the universe as an integrated whole in which the supernatural realm could be understood, at least partially, through observations of events in the natural world. Humans could speculate about God, angels, the devil, and demons because their actions were similar to human actions, although qualitatively different. Aquinas further believed that sensory experience presented reliable information about the actual world, at least if confirmed by a majority of people, because God would not have allowed men to be chronically mistaken. This led Aquinas to accept the existence of such beings as Satyrs, Fauns, and incubi because "many persons" attest that they have seen them. "Hence it seems folly to deny it."[69] Given this epistemological framework, Broedel contends that Kramer and Sprenger's feminization of witchcraft was neither extreme nor radical in itself because it simply confirmed popular conceptions. What was radical, however, was their rationale for why women were witches, a rationale that led them to compile "a veritable *summa* of late-medieval misogynist commonplaces."[70]

The question of exactly what constituted sound and reliable evidence runs though the writings of inquisitors and demonologists. Following Aquinas, Inquisitors like Kramer and Sprenger were convinced that the experiences described by witches (under torture, to be sure) provided sufficiently valid evidence. Witches were the expert witnesses, whose testimony affirmed that demonic copulation was not a figment of their overactive imagination, a thought they voice, revealing their own doubts: "The theory that modern witches are tainted with this sort of diabolic filthiness does not depend so much on our own opinion, as on the expert testimony of the witches themselves, which has made all these things credible."[71] The problem with this line of reasoning is twofold. It raised the issue that would become central in the seventeenth century of whether experience based on the human senses can provide accurate knowledge of external reality; and it brought into question the value of second-hand testimony. Both issues come to the fore in Gianfrancesco Pico della Mirandola's treatises on witchcraft. Pico was a nephew of Giovanni Pico della Mirnadola, but he shared none of his uncle's youthful optimism or heterodox ideas. He was one of the first to employ the skepticism of Sextus Empiricus to undermine philosophy to support faith.[72] In his treatise on witches, *Strix* (1523), a dialogue between an Inquisitor, a skeptic, and a moderator (himself), Pico's nephew takes travel reports as an example. The Inquisitor asks the skeptic if he believes what he hears about crossing the Atlantic or arriving in the gulf of India. When the skeptic admits he does, the inquisitor is quick to point out that there is no difference between the accounts of travelers and those describing witches and the Sabbath; both are secondhand. Their truth lies in the credibility of the narrator. The skeptic is convinced and agrees with the moderator: "when many people are of the same opinion about something, and agree about it as if speaking with one voice, it cannot seem credible that someone goes on claiming the right to deny it."[73] Thus for the younger Pico the truth of witchcraft lay in narratives given by trustworthy people who have witnessed the activities of witches.

A similar conclusion was reached by Bartolomeo Spina, who argues that human society would cease to exist if we only believed what we have ourselves experienced.[74] Steven Shapin and Simon Schaffer describe the same process of witnessing and emphasis on the probity of the witnesses as the process through which Boyle's chemistry and the "new" science were validated.[75] One might object, and scholars have, that the Scientific Revolution was validated by more than just witnessing, that a model of experimental research was put in place that depended on the successful repetition of experiments. Interestingly enough, in some cases witch theorists also took an experimental approach, describing experiments in regard to such things as the ointments that supposedly enabled witches to fly to Sabbaths, an experiment, which, as we have seen, was performed by della Porta. But unlike della Porta, who concluded that the woman in question was delusional as a result of narcotics, demonologists took the fact that the experiments failed as proof that the activities of witches were unnatural because they were demonic. In other words, the ointment used by witches had no natural inherent power; demons actually did the flying. This may not seem to us to be a valid scientific conclusion, but it does reveal that witch theorists were involved in the scientific debates of their time.

Throughout history women have always been vulnerable scapegoats during periods of social and political unrest. From the time of the elder Cato, who blamed the decline of the Roman Republic on the lasciviousness and greed of Roman women, to present-day conservative religious movements, women have been cast in the role of human barometers registering the moral climate of nations. For men, chaste, silent, and submissive women are synonymous with a strong, well-ordered state. The Anabaptists at Munster proclaimed the death penalty for wives who were insubordinate.[76] This was one of the admittedly more extreme solutions to the very real problems in the early modern period. Warfare, religious conflict, social unrest, and the explosion of new information undermined established authority in the political, social, and cultural realms and contributed to the growth of skepticism. A new system of order was desperately needed, and in major respects it was built on the backs of women, especially witches. What demonologists like Kramer and Sprenger were essentially doing was offering their contemporaries a new conceptual framework with which to view the world, an ordering system, so to speak, which explained evil in terms of female sexual sins and attributed social disorder to disorderly women. Broedel makes this point:

> . . . through the creation of an ordered semantic and intellectual system, Institoris [Kramer] and Sprenger provided the necessary terms of

> a satisfactory symbolic discussion of human sexuality, order, and power. In this new conceptual field, disordered sexuality is identified with the devil, inverted gender roles and sexual dysfunction with witchcraft, and defective social and political hierarchies with women and women's sins. None of this, however, is possible without the use of witches and witchcraft as an ordering term; witchcraft, as it were, provides the conceptual grid which binds this cognitive map together.[77]

If we take Broedel's notion of witchcraft as an "ordering system" and combine it with Stephens's claim that demonologists provided anxious early modern males with a "new theodicy," we can see that the witch hunts of the early modern period were not aberrations in an otherwise progressive age, but an essential element in the creation of a new world order based on the elimination of dangerous and polluting female sexuality.

As Mary Douglas and other scholars have argued, every society's cosmology includes some kind symbolic connection between pollution and danger along with injunctions about how to eliminate both and restore purity. Even such mundane things as spring cleaning represents an attempt on a personal level to re-order the environment by exposing and eliminating the dirt that is literally under the rug. On a cosmic level sacrifice has been one of the most prevalent ways to restore purity and rid the environment of disorder and the dangers it unleashes. In early modern Europe witches were forced to assume the role of sacrificial scapegoats. Their elimination would restore social equilibrium and eradicate pollution. Witches represented the demonic forces that threatened to erupt from the underworld engulfing those above. According to Broedel, as a result of the publication of the *Malleus maleficarum* this view of the witch held true for both learned and popular culture. Thus, however much early Protestants gave lip service to the spiritual equality of men and women and to the important role women should play in the household, this did not manifest itself in actual practice, at least in the short term. A new "theology of gender" emerged that became a key component in the creation of a new world order based on the formation of a "Godly State," the subject of the next chapter. This "theology of gender" affected Protestants and Catholics alike. It subordinated women to men as never before and constricted their roles in the public as well as the private sphere in new and unprecedented ways.[78] Witch hunting played a crucial role in these developments.

CHAPTER 5

THE "GODLY STATE"

Max Weber was mistaken in thinking that rationalization and modernization were a primarily Protestant phenomenon, for there were close parallels between developments within Protestantism and Catholicism in the course of the European Reformations. Haunted by the disorder accompanying the breakdown of religious consensus, church and secular authorities on both sides of the religious divide joined forces to establish efficient and stable bureaucratic societies grounded in an obedient, disciplined, and orthodox citizenry, whose primary allegiance was to church and state.[1] What modern scholarship has increasingly shown is that the great age of faith did not occur, as so many people mistakenly believe, in the Middle Ages but in the early modern period, when a concerted effort was made by authorities to create "Godly" states on the backs of upright, God-fearing citizens. As Robin Briggs says, "Godly discipline was the theme of the age, for Catholics and Protestants alike."[2] We can see this in the kinds of names English Puritans gave their children: Discipline Whiting, Arise Evans, Sobriety Bollsby, Wisdom Hampon, and Contented Bird.[3] Medieval Christianity had been more realistic, expecting holiness from the relatively few individuals who chose a monastic life and sin and repentance from the rest. Early modern authorities, however, demanded holiness across the board as they applied traditional Christian ethics to the population as a whole in much stricter ways than ever before. A concerted effort was made to reform peoples' behavior as well as their beliefs in what Norbert Elias famously described as the "Civilizing Process."[4] An essential element in the campaign to reform morals that was central to this process was the imposition of strict boundaries. Everyone was to know his or her proper place in the moral and physical universe. Lyndal Roper's description of the morals campaign among German Protestants holds just as true for Catholics:

> The morals campaign. . . . proposed a world where each individual had duties according to his or her place, the master, mistress, children and servants, each having a particular set of moral obligations. . . . Order and right living were its catchwords. It called for punishment of the sins of drunkenness, evil living and Godlessness, and it exhorted secular authority to act to bring the Godly society into being.[5]

The concern with establishing boundaries was an important aspect of the emphasis on contraries described in earlier chapters. Mary Douglas argues that differentiating and demarcating one thing from another is a basic way of imposing order on experience: ". . . ideas about separating, purifying, demarcating and punishing transgression have as their main function to impose system on an inherently untidy experience. It is only by exaggerating the difference between within and without, above and below, male and female, with and against, that a semblance of order is created."[6]

The concern with establishing a new order that was such a fundamental feature of early modern thought is reflected in the palpable horror of mixing things indiscriminately. In his biography of Calvin, William Bouwsma writes that Calvin "abominated" mixture; the very word was "one of the most pejorative terms in his vocabulary."[7] For Calvin mixture was synonymous with "adulteration," and "promiscuity," and ultimately with the disorder and confusion he saw around him and tried to remedy. Clearly, Descartes was not alone in craving clear and distinct ideas. Bouswma attributes the clarity of Calvin's style to this concern: "he stabilized the meanings of words . . . but therefore also the structure of the universe he inhabited, by such linguistic devices as frugality in the use of adjectives."[8] Because boundaries kept mixture in check, Calvin was obsessed with them. He condemned curiosity precisely because it encouraged men to "greedily overstep their boundaries to inquire into the truth."[9] Just as God had established boundaries to what could be known, so had he established boundaries between peoples: ". . . men are placed on the earth so that each nation may be content with its own boundaries."[10] Gender boundaries were of paramount importance. Men who were effeminate and women who affected "manliness in their dress and gestures" were a disgrace. Special targets of Calvin's contempt were women who resembled "lansquenets" (German mercenary soldiers) and who aspired to shoot "as boldly as a man." These women were "monsters so scandalous that one ought not only to spit at meeting one but pick up some piece of filth to throw at them for so audaciously perverting the order of nature."[11] The idea of throwing excrement at these women indicates the depth of Calvin's distaste, for, as Bouwsma points out, excrement represented to Calvin the horror of everything that was unformed and without limits, in other words, everything without clear and distinct boundaries: "Excrement, for Calvin, was not simply matter out of place: as an image for formlessness, that is of chaos, it stirred up his deepest horror of nonbeing."[12] Calvin associated excrement with impurity, pollution, and contagion, all of which were summed up by sin. Inasmuch as humans are innately and ineradicably sinful, Calvin claimed that "We take nothing from the womb but pure filth. . . . it is certain that there is no one who is not covered with infinite filth." The doctrines of Rome were nothing less than "stinking excrement."[13]

To understand why establishing boundaries was such a crucial aspect of early modern life it is useful to turn to Elias and Douglas. Both have provided insights into the way dirt was and is connected with disorder and cleanliness with purity and order. Elias correlates new concepts of personal hygiene with changing political and social structures that placed increasing psychological and social constraints on individuals. Douglas takes a more theoretical approach: first, by showing that the very concept of dirt implies a system in which impurity is synonymous with disorder and danger and, second, by emphasizing the way the pollution that comes from dirt involves crossing physical boundaries, be they the boundaries around an altar, a temple, an individual body, or the body politic.[14] For early modern Europeans the holiness of the body was crucial, and in this regard the homonym in English between holiness and wholeness hits the nail on the head. In order for an individual's body to be holy it had to be whole. Calvin, like many Christians before and after, saw heresy and disbelief as an infectious contagion threatening the integrity of the true believer: "Nothing is more infectious than association with the unGodly, for, since we are inclined to vice, it cannot be but that, when we frequent corruption, the contagion spreads more widely." Douglas considers the fear of contagion and concern with bodily integrity the basis of pollution rituals:

> The body is a model which can stand for any bounded system. Its boundaries can represent any boundaries which are threatened or precarious. . . . We cannot possibly interpret rituals concerning excreta, breast milk, saliva and the rest unless we are prepared to see the body as a symbol of society, and to see the powers and dangers credited to social structure reproduced in small on the human body.[15]

In post-Reformation Europe, the connection between holiness and wholeness applied not only to the physical body of believers but to the body of scripture and the body politic as well. The porosity of the boundaries of all three had to be carefully guarded from the intrusion of external and dangerous entities such as witches, heretics, and the slew of demonic forces poised to snare the unwary. But the threat did not come solely from the outside. What is new and remarkable in the early modern period is that danger came from within as well in unprecedented ways, from the insidious and menacing physical and emotional drives within individuals that made them want to cannibalize and incorporate everything beyond themselves from food to foreign lands, riches, and people.[16] These were so unpredictable and uncontrollable that they required constant monitoring, discipline, and restraint. Self-scrutiny became a defining characteristic of the age of Reformations. Both Catholics and Protestants were exhorted to examine themselves for the smallest infraction. As one Catholic author advised, the reader should "proceed . . . with shame, to examine himself, as if he were a criminal before his judge, from whom he expects a death sentence."[17] "Depraved," "corrupt," "sinful," and "filthy" are words that continually appear in evangelical discussions of human nature. According to Jonathan Edwards (1703–1758), sinful men

> are totally corrupt . . . and in all their dispositions and affections, their heads, their hearts, are totally depraved, all the members of their bodies are only instruments of sin; and all their senses, seeing, hearing, tasting, etc., are only inlets and outlets of sin, channels of corruption.[18]

Cotton Mather lamented his condition: "I have certainly been one of the filthiest creatures upon Earth."[19] The itinerant preacher Daniel Rogers took self-loathing to even greater heights: "The Lord was pleased to give me a farther sense of the hateful Nature of sin from a view of the working of It in my own Heart. So that I did sensibly loath and abhorr It as the vilest filthiest Thing in all the world—I hate It with a perfect Hatred—It is nauseous."[20] New forms of religious devotion and ritual emerged to accommodate this emphasis on discerning and repenting for inward sinfulness. Frequent confession was advised for Catholics, along with solitary meditation in private chapels and bedrooms equipped with specially designed furniture, the prie-dieu. In his instructions to confessors Charles Borromeo recommends that priests keep a chart describing the state of the souls of each parishioner.[21] New categories of sin were elaborated among Catholics, and the proliferation of religious confraternities and Marian congregations had the express purpose of transforming "men from top to bottom."[22] As Louis Chatellier puts it, "Purity of self, purity in others; these were the aims."[23] Nothing was to be hidden or withheld. Speech was to be plain and transparent with no rhetorical flourishes: "Above all, one should always . . . express our thoughts with great simplicity and in a language redolent of the Gospel, without affecting certain thoughts which are only curious and without choosing words which might distract or detain the spirit. . . . "[24] Michel Certeau perceptively likens this practice of ferreting out secret sins to witch hunts in the larger society: "The proliferation of religious confraternities and congregations contributed to that crusade in which the hunt for secrets took on the appearance of a witchhunt."[25] With fewer church institutions to help them in their self-examinations, the private diaries kept by many Protestants allowed them to record their innermost thoughts in excruciating detail and take stock of their spiritual state. But for both Catholics and Protestants the spiritual state was unfortunately dependent on those physical urges that seemed determined to overwhelm every good and moral intention.

The connection between sin and disease runs like a leitmotif through Christian history and contributed to the morbid disgust for the body that emerged with special force in the early modern period. The belching, farting body that was a source of humor in medieval fabliaux as well as Rabelais becomes a serious source of embarrassment in the Reformation period (and forever after). The civilizing process involved the repression of normal physical needs in the interests of decorum and propriety. Bodies were increasingly separated from other bodies and cordoned off into protected zones. The individual bed and individual tomb came into fashion in the eighteenth century, and both were examples of the kind of new boundaries that were being erected to isolate individuals from one another. As Laporte succinctly says, "To each his cesspool and to each his grave."[26] There was less touching and embracing. Florid gestures like prostrating oneself at the feet of a person of authority (or a woman one wanted to impress) gave way to discreet, understated gestures. A new modesty prevailed that viewed the codpiece with distaste and an uncovered breast with disapproval: "Cover this breast, which I do not like to see," orders Tartuffe.[27] Wedding guests no longer put newlyweds to bed and returned in the morning to check the sheets. New forms of architecture separated private living quarters (with bedrooms and toilet facilities) from public rooms. Defecating in public was now looked upon with repugnance.[28] A palpable disgust for the body is a constant theme among evangelical Protestants. Cotton Mather thought he was no better than a dog because he urinated: "I was once emptying the *Cistern of Nature*, and making *Water* at the wall. At the same Time, there came a Dog, who did so too, before me. Thought I; 'what mean, and vile Things, are the Children of Men, in this mortal State! How much do our natural Necessities abase us, and place us in the same regard, on the same Level with the very Dogs!' "[29] Jonathan Edwards was equally appalled by his body: "The inside of the body of man is full of filthiness, contains his bowels that are full of dung, which represents the corruption and filthiness that the heart of man is naturally full of."[30] Disciplining recalcitrant flesh became a key aspect of the civilizing process. New disciplines for self-improvement emerged, involving dieting, exercise, hygiene, lessons in deportment, and the avoidance of "self-pollution." The threshold of embarrassment and shame was raised exponentially as the compulsion emerged to hide, deodorize, cleanse, purge, and expel every disgraceful aspect of the physical self.

In his entertaining, hugely informative, and unfortunately last book, *Flesh in the Age of Reason*, Roy Porter describes the repugnance people felt for their bodies,

> To a degree hard to imagine nowadays, visible, tangible flesh was all too often experienced as ugly, nasty, and decaying, bitten by bugs and beset by sores; it was rank, foul and dysfunctional; for all of medicines' best efforts, it was frequently racked with pain, disability and disease; and death might well be nigh.[31]

As we have seen, this repugnance for the body began two centuries earlier and became an integral part of Reformation theology. The obsession with controlling the body and especially bodily orifices provides the subject matter of "Grobian literature," which originated with the publication of Friedrich Dedekind's ironic poem *Grobianus et Grobiana, De morum simplicitate, libri tres* (1549). Dedekind and the Grobian authors who followed him sought to civilize males by inculcating the idea that boorish, coarse behavior is unmanly and, even more to the point, a sign of effeminacy. Grobian literature obsessively focuses on the materiality of the body, on its uncontrolled orifices as they vomit, excrete, urinate, and exude snot, semen, and blood. The object of this literature is to shame male readers into proper, civilized behavior. This consists in subjugating and sanitizing the delinquent body, which is by definition female.[32] The connection of women with the body and matter and men with the soul or spirit goes back to Aristotle and appears as a constant theme in the literature of the Middle Ages.[33] In the early modern period this dichotomy was drawn in starker terms than ever before. As Pierre Bourdieu observes, the project of civilizing the body is literally embodied in such things as dress, comportment, and verbal and physical mannerisms. By these means what is essentially a purely social construct of gender identity is made to seem "real" and to reflect the divinely ordained natures of males and females.[34]

Scholars have commented on how repulsive and disgusting Grobian literature is when it comes to sex; and inasmuch as sex is associated with women, women are equally disgusting and unappetizing. Grobian literature was written for men, and even though reference is made to a female Grobiana, she is in an entirely different category. Grobianus is not by nature a boor; he is only a boor because he chooses to succumb to female influence. Grobiana, however, is what she is because she is inescapably female. As Correll comments, "Presumably you can take the Grobiana out of Grobianus and produce the civil subject that is, in fact, the project of the civilizing process . . . but . . . you cannot take the Grobiana out of Grobiana."[35] The upshot is that in the case of women shame and self-discipline are not sufficient to civilize them; they must also be sequestered and excluded from the public realm not only for their own good but especially for the good of men.[36] This brings us back to the way gender became a means to organize society and create stability in the early modern world. As we have seen, one of the most salient characteristics of witches was the fact that they were either unmarried or unhappily married. This underlines the new and problematic role that marriage and gender roles within marriage assumed in the early modern world.

Lyndal Roper and Merry Wiesner-Hanks have been in the forefront of scholars who argue that the institution of marriage and the regulation of gender relationships within marriage became a central concern in the Reformation and post-Reformation world as theologians, philosophers, scholars, and statesmen came to grips with the realities accompanying the rise of the middle class, a new and unprecedented degree of social mobility, and the disruptive role of money in societies whose structures were still essentially feudal.[37] During the Middle Ages society was "diffusely familial."[38] Given the weak power of the state, individuals had to rely on the family for protection and economic security. Consequently, the concept of kinship was far more inclusive, allowing for different degrees of relatedness involving various levels of reciprocal obligation. Both Protestant and Catholic reformers as well as governmental authorities were determined to minimize these diffuse "family" associations and replace them with direct links between individual nuclear families and religious and secular authorities. In their campaign to reform the manners and morals of the laity, Protestants emphasized the nuclear family as the basic social unit and created ecclesiastical institutions and courts to oversee family life, while reinforcing patriarchal authority in the home.[39] Calvin's doctrine of the utter depravity of fallen human nature led him to champion the state and family as bulwarks against chaos. As Michael Walzer comments, "the permanent, inescapable estrangement of man from God is the starting point of Calvin's politics." For Calvin, fearfulness, anxiety, distrust, and war characterized the human condition—which was, indeed, the case in war-torn, early modern Europe—and the only solution was the rigid enforcement of authority predicated on repression and obedience.[40] Calvin transformed the institution of fatherhood into a religious office, stressing patriarchal authority over affection.[41] Reformed Catholics were no different from Protestants in this regard. They devised a new language of devotion that focused on the royal and paternal aspects of God, God as father, master, king, and judge. Both groups presented God as an "extraordinarily demanding, almost domineering father figure," who required utter dependence and obedience from his "sons."[42]

The paradox of Protestantism lies in just this: how did a radical evangelicalism predicated on the spiritual equality of all Christians, irrespective of gender, became a bulwark of a hierarchical social order solidly built on inequalities of gender, inequalities that remain alive and well among conservative Protestants to this day.[43] Roper contends that the new Protestant ideology of marriage lies at the heart of this reversal inasmuch as it promoted an agenda of order and discipline predicated on the sole authority of husbands and fathers in the family and of male authorities in society at large.[44] Thus, in both theory and practice there was little to distinguish Protestants from Catholics when it came to gender roles in marriage. Patriarchal ideology was reinforced in both groups to such an extent that the father became the "legalized petty tyrant within the home."[45] One might well ask why marriage became such a central issue in the mind of reformers. The short answer is sex and sin. Following Augustine, Luther and Calvin rejected the relative optimism about human nature characteristic of Renaissance Humanists and emphasized instead the utter depravity of human beings. And like Augustine, they envisioned the nature of this depravity largely in sexual terms. To quote Calvin:

> . . . our nature is not only destitute of all good, but is so fertile of every evil that it cannot be idle. Those who have said that original sin is "concupiscence" have used an appropriate word, if only it be added—something that most will by no means concede—that whatever is in man, from the understanding to the will, from the soul even to the flesh, has been defiled and crammed with this concupiscence. Or, to put it more briefly, the whole man is of himself nothing but concupiscence.[46]

The only institution that could contain this concupiscence, at least to some degree, was marriage, but not marriage in the old Catholic sense, only marriage in a new Protestant form, in which gender roles and sexuality were elaborately defined and surveillance mechanisms established to make sure that husbands and wives behave appropriately. The draconian nature of the kind of surveillance envisioned by Protestants is illustrated by the instructions issued to "The Office of Elders" in Geneva. These stipulate that the Elders examine not only the behavior and speech of the citizens but also observe their attitude and intentions by visiting every household at least once a year and interrogating the residents. As the instructions ominously say, ". . . in every precinct a lay elder should be chosen, who 'can have his eye everywhere.' "[47] Roper describes the all-embracing nature of the "Discipline Ordinance" promulgated in Augsburg in 1537 with its attempt to micromanage the rights, duties, and appropriate relationships of individual family members:

> The sexual discipline which the whole citizenry was to adopt was both more all-embracing and less well defined than it had been before the Reformation. Now any sexual relationship outside marriage was counted sinful and any occasion on which the sexes mingled, such as dances, might lead to sin. So absolute were the demands of the ideal that the Council was drawn inevitably to define marriage and the relations which ought to hold between husband and wife, parents and children, masters and servants as it articulated the concept of discipline.[48]

What is so compelling about the concupiscence these regulations were established to monitor and restrain is that it was considered basically a female problem.[49]

Among Protestants, household duties and child-bearing became the acceptable limits of female horizons. Luther made this abundantly clear when he likened the good wife to a "snail" or "a nail driven into the wall" while commenting on the divinely ordained difference between the roles of husbands and wives:

> The rule remains with the husband, and the wife is compelled to obey him by God's command. He rules the home and the state, wages war, defends his possessions, tills the soil, builds, plants, etc. The woman on the other hand is like a nail driven into the wall. She sits at home . . . the wife should stay at home and look after the affairs of the household as one who has been deprived of the ability of administering those affairs that are outside and concern the state. . . . In this way Eve is punished.[50]

The fact that women were created with "broad hips and a wide fundament to sit upon" clinched the case for Luther inasmuch as the physical proof that women were to "stay at home," "keep house, and bear and raise children" was inscribed in their anatomy.[51] When it came to defining the separate spheres appropriate for husbands and wives, Catholic and Protestant males were joined at the hip. There is little to distinguish what Luther said above from the Catholic author of *La famille saincte*, who, like Luther, minced no words on this topic: "I find two sorts of occupation in a family, the one requires the husband, the other the wife. External matters are more proper for the husband; internal matters are the business of the wife." The wife must stay out of her husband's business: ". . . it is intolerable for a woman with a husband who is not a fool to get involved in business affairs. . . . Women should be like the soldiers of a garrison, who are never permitted to go beyond the walls to do battle."

It was not only the biblical punishment of Eve that mandated women's primary role as the bearers of children, but the science of the day as well. At the beginning of the period the prevailing view was that male and female sex organs were mirror images of each other. This was Vesalius's opinion. As his student Baldasar Heseler, wrote: "The organs of procreation are the same in the male and the female . . . For if you turn the scrotum, the testicles and the penis inside out you will have all the genital organs of the female."[52] Michael Stollberg and others have shown, however, the idea that the sex organs of men and women were radically different came to the fore at the beginning of the seventeenth century as the idea of gender complementarity emerged. The notion of complementarity was predicated on the idea that women's unique function, marriage and motherhood, was ordained by their anatomy. By their very nature women were consequently excluded from everything that was culturally and intellectually valued in the public world of men.[53] The emphasis on women's role in the family occurred at the same time their economic opportunities narrowed. Women, to use Merry Wiesner's phrase, were "domesticated," and the split between the public and private sphere so marked in the gender discourse of the nineteenth century was well in place by the end of the seventeenth.[54]

Even in their prescribed role as wives and mothers, women remained troublesome enough to be banished from much of the historical and religious record. Although Luther retained great respect for the Virgin Mary, he was appalled by what he considered Catholic idolization of her. Luther called Mary by the title "Theotokos" (Mother of God), but he rejected the active invocation of Mary formulated in prayers such as the "Hail Mary." The ability of Mary to intercede with God and Christ for sinners was to give her far too much power:

> Furthermore, how will you endure [the Romanists'] terrible idolatries? It was not enough that they venerated the saints and praised

> God in them, but they actually made them into Gods. They put that noble child, the mother Mary, right into the place of Christ. They fashioned Christ into a judge and thus devised a tyrant for anguished consciences, so that all comfort and confidence was transferred from Christ to Mary, and then everyone turned from Christ to his particular saint. Can anyone deny this? Is it not true?[55]

With the abolition of saints, including female ones, and the demotion of Mary to a suitably subservient position, Protestant women were deprived of female role models other than that of an obedient wife; and with the disestablishment of monasteries, they had no satisfactory alternative to marriage. But even Catholic women lost the positive female role models they once had as Catholic male reformers, just like their Protestant counterparts, transferred the gentle, nurturing qualities previously associated with the Virgin and female saints to God, the Archangel Michael, and St. Peter. Thus during the early modern period, whether in heaven or on earth, males assumed whatever few qualities they considered positive about women, thus marginalizing them in the one role for which they had ostensibly been designed. As Reinburg comments, "The devotional universe of Protestant and Catholic clerics was governed by a clear, simple, male hierarchy that ruled with sovereign authority."[56]

A similar appropriation of female nurturing powers occurred among rulers intent on consolidating their power. King James I of England envisioned himself as the single parent of his realm, as "loving nourish-father" who provided his subject with "their own nourish-milk."[57] Robert Filmer, whose *Patriarcha* argued for the divine right of kings, claims that kings are like fathers because both are the sole parent in their respective realms: "We find in the Decalogue that the law which enjoins obedience to Kings is delivered in terms of: Honour thy Father."[58] Filmer simply leaves out the rest of the commandment, "and thy mother." He was not alone. Thomas Hobbes excludes the mother from his definition of the family: ". . . a great Family if it be not part of some Common-wealth, is of it self, as to the Rights of Soveraignty, a little Monarchy; whether that Family consist of a man and his children; or a man and his servants; or a man, and his children, and servants together; wherein the Father or Master is the Soverign."[59] For Protestants the husband and father assumed the role of God, priest, and ruler in the "little commonwealth" of his family. Catholic women could at least go a priest when things got tough at home;

but for Protestant women in a very real sense the priest, pope, and king lived at home.

The male appropriation of the female role in parenting helps to explain why positive statements extolling mothers and motherhood are noticeably absent in texts dealing with conduct, just as mothers are noticeably absent in literature and art. Literary scholars have called attention to how often mothers were left out of literature, autobiography, and family portraits in the sixteenth and seventeenth centuries and how common it was for men to assume the female procreative role.[60] Louis Montrose comments on the significant lack of mothers in Shakespeare.[61] Mothers are absent from male autobiographies as well. One would hardly know from reading the autobiographies of Richard Baxter and John Locke that they were of women born.[62] Jonathan Goldberg notes the same omission of mothers in Stuart family portraits.[63] Even where wives and mothers are portrayed, there are subtle visual codes emphasizing patriarchal authority. Husbands are on the left of the picture plane so that the viewer would see him first. The male often gestures toward his wife, reinforcing the idea of her subordination, and she is depicted in a passive, demure manner.[64] Given the evidence it is impossible to accept Steven Ozment's contention that Reformation morality allowed women "a position of high authority [as mothers] and equal respect [to men]."[65]

Scientific texts dealing with reproduction exhibit the same marginalization, even elimination, of women. A pseudo-Paracelsian work described menstrual blood as the poisonous matrix from which monsters like the basilisk originate,[66] a notion consistent with Paracelsus's undisputed statements to the effect that women were the source of all evil.[67] Given this conviction together with the Aristotelian "scientific" notion that the father is the only true progenitor of offspring since the mother merely acts as an incubator for the male "seed," it is understandable that Paracelsus and many alchemists and natural philosophers devoted their energies to producing *homunculi*, marvelous creatures created by males exclusively from male semen and consequently uncontaminated by female characteristics.[68] The elimination of women is also a standard feature in alchemical texts. Carl Jung's view of alchemy as a psychic and spiritual process in which the male and female aspects of the individual are unified in a psychologically satisfying whole symbolized by the image of the hermaphrodite simply does not describe what alchemists thought or did.[69] Although the philosopher's stone is frequently described as the offspring of a "royal" couple, in many cases this birth is better described as an act of cloning because the philosopher's stone emerges solely from the father. This is the message delivered by Lambspring (or Lambsprinck) in his book of alchemical emblems, in which the king swallows and then gives birth to his son entirely on his own without his wife or any other woman present.[70] A similar message appears in the work of the German Lutheran alchemist, physician, and self-styled poet laureate, Michael Maier. In Maier's alchemy, male and female are polar opposites, and they remain polar opposites. When males need the one power with which females are credited, namely the ability to bear offspring, they simply appropriate it, making women superfluous. The first and second emblems in Maier's *Atalanta Fugiens* (1618) illustrate this. The first depicts Hermes Trismegistus, the reputed founder of alchemy in the West, as pregnant with the unborn child. Emblem 23 takes this step of an exclusively male act of generation to its logical conclusion, depicting Zeus giving birth to Athena fully formed from his head. Zeus is, of course, in good company because according to the creation story in Genesis Eve was "born" from Adam, or at least from Adam's rib, a "fact" reiterated by Paul and Christian theologians ever after. Maier's second emblem underscores this reversal of roles. Here the mother is relegated to the passive role of nurse, as she had been by Aristotle, caring for but having no active role in the procreation of the child. An even stronger repudiation of the maternal role appears in emblem 49, which depicts five men standing around an ox hide, into which, to put it delicately, they deposit their semen, from which a fully formed male will presently develop. Two further emblems offer more sinister scenarios because they make it clear that when females are involved in procreation, even if just as incubators and nurturers, their role is short-lived because their death is the prerequisite to a successful "birth." In emblem 5 a handsomely dressed courtier carrying a spotted toad approaches an elegant lady on a deserted city street. The courtier suddenly thrusts the toad onto the woman's breast. With an expression of appreciable horror, the woman watches as the toad begins to nurse. The accompanying epigrammatic verse explains the scene:

> To woman's breast apply the chilly toad,
> So that it drinks her milk, just like a child.
> Then let it swell into a massive growth,
> And let the woman sicken, and then die.

Michael Maier, *Atalanta Fugiens* (Oppenheimii, 1617). Emblem 49 (The Fathers of Orion). (Wellcome Library, London)

> You make from this a noble medicine.
> Which drives the poison from the human heart.[71]

The final and fiftieth emblem depicts a woman lying in a shallow grave with a serpent coiled around her body. This is even more gruesome and perverse in its association of the grave with the marriage-bed:

> A deep grave's dug for Dragon venomous,
> Who's with the woman tightly intertwined.
> While it enjoys its marriage-bed, she dies,
> And with the Dragon's buried in the ground.
> Thereby its body dies, suffused with gore,
> And this is the true pathway of your work.[72]

These images belie the idea that alchemists championed the "view of nature and woman as Godly," to quote Evelyn Fox Keller.[73] In alchemy

Michael Maier, *Atalanta Fugiens* (Oppenheimii, 1617). Emblem 5 (Courtier, woman, and toad). (Wellcome Library, London)

female nature is routinely tortured, dismembered, sacrificed, drowned, dissolved, and devoured in ways worthy of the Marquis de Sade. Far from cherishing female nature these engravings show that alchemists resuscitated the Aristotelian notion that the form or essential nature of the child is contained in the male semen and that the female is nothing more than an incubator. This doctrine had been challenged by Galen, but with the triumph of the new science it reemerged, and we find scientists who claim to have made microscopic observations of spermatozoa containing perfect miniature fetuses.[74]

The essentialist rhetoric that developed about sex and sexuality in the early modern period undermines another supposedly positive aspect of alchemy, namely the prominence of hermaphrodites in alchemical imagery. Hermaphroditism became an important issue in the sixteenth

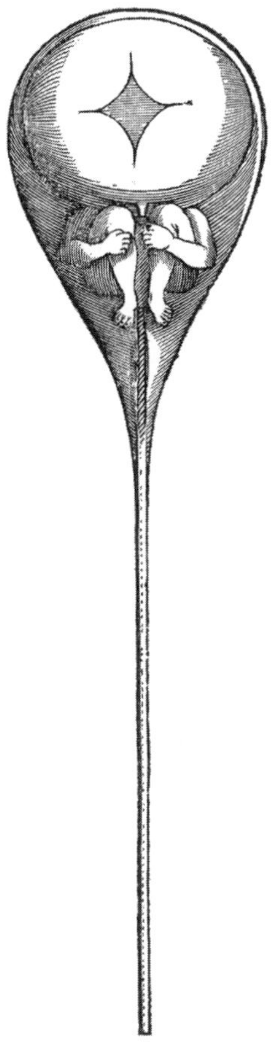

Spermatozoon/Homunculus. (From Nicolas Hartsoeker, *Essay de dioptrique*. Paris, 1694. Wellcome Library, London.)

and seventeenth centuries because the hermaphrodite was both male and female and therefore did not fit the increasingly rigid gender stereotyping that came into force.[75] Although some scholars, notably Carl Jung, Mircea Eliade, and Edgar Wind, followed by Carolyn Merchant and Evelyn Fox Keller, consider the image of the hermaphrodite an affirmation of gender equality, more recent studies suggest this is not the case.[76] Raymond Waddington, for example, rejects Wind's positive evaluation of Niccolò Bellin da Modena's depiction of Francis I as a hermaphrodite and sees it instead as a sophisticated joke deflating Francis's pretensions to both military and sexual prowess.[77]

The idea that images of hermaphrodites or discussions of androgyny offered a positive alternative to notions of gender polarity overlooks the fact that the concept of androgyny implies that males and females are polar opposites.[78] Any suggestion to the contrary aroused fears. Lorraine Daston and Katherine Park point out that in French civil law an individual's gender determined his or her capacity to marry, inherit, act as a witness, assume the position of guardian, and hold political office. The tremendous weight accorded to the question of a person's gender meant that "sexual ambiguity was . . . not legally tolerated."[79] The anxiety caused by hermaphrodites expressed a more general and intense fear of homosexual acts as well. However prevalent homosexuality may have been in court circles, the practice was generally viewed with revulsion and horror because it blurred the line between the sexes by undermining the dichotomy between the "active" male and "passive" female. In this context it is understandable that Amazons became a subject of considerable interest in the early modern period, for with their military skills and refusal to marry they too blurred increasingly rigid gender stereotypes.[80] Maryanne Horowitz claims that playing with notions of gender was an acceptable Renaissance pursuit, but she admits that however playful various authors may have been in their discussions of autonomous female figures, they successfully refashioned them along increasingly polarized gender lines:

> . . . in the Renaissance cross-cultural questioning of gender categories, authors and artists enjoy playing with Amazons and other ancient figures who transcend accepted gender stereotypes. An apparently amusing "male" game, presented for the enjoyment of conventional men and women (as in Shakespeare's *Taming of the Shrew*), is refashioning autonomous female figures along traditional gender lines, that is, appropriating them to conserve the status quo.[81]

One wonders if this kind of literary imaginative activity can really be described as "playing," especially since, as Horowitz admits, these texts end up by assuring "those frightened by female independence that even an Amazoncan become a conventional woman."[82]

The fear of independent and overbearing women affected Catholics as well as Protestants, but there were inherent contradictions in the Protestant attitude toward women that made women especially bothersome.

While women were admitted to be men's spiritual equals on the basis of Galatians 3:28 and worthy of love and a position of authority in the household, they were declared men's inferiors in every other respect. The tortured way that Protestant males deal with this discrepancy comes out in the following passages written by the English scholar William Tyndale (c. 1494–1536), who made one of the first translations of the Bible into English:

> Ye must understand that there be two states or degrees in this world: the kingdom of heaven, which is the regiment of the gospel, and the kingdom of this world, which is the temporal regiment. In the first state there is neither father, mother, son, daughter, neither master, mistress, maid, man-servant, nor husband, nor wife, nor lord, nor subject, nor man, nor women: but Christ is all. . . . In the temporal regiment, thou art a person in respect of others; thou are an husband, father, mother, master, mistress, lord, ruler, or wife, son, daughter, servant, subject, etc. And there thou must do according to thine office. If thou be a father, thou must do the office of a father, and rule, or else thou damnest thyself. Thou must bring all under obedience, whether by fair means or foul. Thou must have obedience of they wife, of they servants, and of thy subjects; and the other must obey.[83]

Tyndale and many Protestants who came after him clearly had a problem explaining exactly why the temporal and spiritual state were so radically different when it came to the position of women, but they never stopped trying. William Whately (1583–1639) simply told women they had to accept this basic fact of life: "[e]very good woman must suffer herselfe to be convinced in judgement, that she is not her husbands equall. Out of place, out of peace; and woe to those miserable aspiring shoulders, which will not content themselves to take their room bellow the head."[84] The writings of Protestant theologians as well as women themselves suggest that Protestant women had taken the teaching of spiritual equality many steps further than male authorities intended. William Gouge (1575–1653), for example, reported that whenever he preached the doctrine of female submission and inferiority, there was a good deal of "squirming" and "murmuring" among the women in the congregation. This simply convinced him that more such preaching was necessary.[85] John Brinsley castigated those women who had "such high and imperious spirits . . . as if they were made only to rule, not at all to obey." What particularly galled him was the fact that women voted with their feet and left his congregation. If they could do that, he warns husbands to think what they might do to them: "[these women] will not stoop to any kinde of subiection, especially to their Husbands. . . . If their Husbands weare the Crown, yet they will sway the Scepter. If their Husbands be in places of Authority, they will Rule with them, if not over them."[86] Although by all accounts Calvin was happily married to a devoted and obedient wife, he believed that all women wanted dominion: "It seems to them that, if they do not have entire mastery, if one does not believe them about everything, if all their advice is not taken, one does them an injury and disdains them."[87]

Gouge, Brinsley, and Calvin had sufficient cause of alarm. Examples of strong, independent women abound in early Protestantism, and these women could match male experts verse for verse in quoting scripture. Like most new religions, Protestants were happy to accept anyone who flocked to their cause, and this included women.[88] Catherine Schülz, the wife of the Strasbourg reformer Matthew Zell, had the temerity to quote Galatians 3:28 and Joel 2:28–29 to the Bishop who had excommunicated her husband: "You remind me that the Apostle Paul told women to be silent in church. I would remind you of the word of this same apostle that in Christ there is no longer male nor female and of the prophecy of Joel: 'I will pour forth my spirit upon all flesh and your sons and your daughter will prophecy'."[89] Argula von Grumbach was another Protestant women who dared to talk back to the male establishment. David Underdown claims that English women working in better-paying trades were becoming more independent economically at the same time that they were granted more spiritual authority in the home. These developments undermined the authority of husbands. Restoring male authority obsessed magistrates, which explains why the number of women prosecuted as scolds by ecclesiastical courts increased.[90] As we have seen, Hugh Peters's condemnation of Anne Hutchison shows how offended male magistrates were by women who transgressed the limits of their allotted sphere. He was much more troubled by her assertive masculine behavior than by her unorthodox religious opinions.

The two prevailing images of women in Protestant thought suggest that husbands were not as secure as they might have wished. The first, the good woman who cheerfully accepts her submissive role as wife and mother, illustrates the way men wished things to be; the second, the "man-kinde" or "masterly wife," a monster or perversion, who had to be tamed and broken like a horse, indicated the way men really viewed the situation. The very real fear that gender relations were changing and getting out of hand was depicted in terms of gender reversal. In a satirical broadside entitled "A Character of a Turn-Coat: Or, the True Picture of an English Monster" (1707), the figures of a man and woman reverse roles when the sheet is turned upside down, an indication that the world is topsy-turvy. The message is explained in a bit of doggerel:

> For as the Times do change, they'll change their Face
> Foreswear their Sex, their Age, their Name, and Race.
> As by these Pictures you may plainly see,
> He that was Man, a Woman seems to be.
> And she that did a Woman represent,
> By change into another form is sent.[91]

The concern with gender was at the core of what Stephen Greenblatt has described as a new preoccupation with the "'self.'"[92] Anxiety about what constitutes the "self" increased during the early modern period as power and class relationships were redefined and reformulated for an increasingly centralized, commercial, and urban society. This is apparent in the case of courtiers and civil servants, for at the heart of both professions lay an inescapable contradiction between the nature of the job and the gender of the job holder. The stance of both groups was essentially feminine inasmuch as their role was to be pleasing, pliant, and subservient. J. R. Woodhouse describes the novelty of this situation and the confusion it occasioned in gender identity. Military prowess and good advice were no longer sufficient to curry favor with a ruler. Wit and the capacity to amuse became the primary avenues to success. But wit calls for the ability to deceive and dissimulate, for talents (if we may call them that) traditionally associated with women. In this situation, the age-old dichotomy between male and female, with all its attendant polarities of active versus passive, dominant versus subordinate, reason versus sense, and public versus private, became problematic, and the issue of what it meant to be male or female assumed a pressing urgency reflected in the increasingly essentialist rhetoric of gender discourse.[93] The fear of being considered effeminate and the stigma attached to it led to a polarization of the sexes that bordered on caricature. We can see this reflected in sixteenth-century fashion with the exaggerated codpieces worn by men and the accent on broad hips characteristic of Elizabethan dresses, both of which became less necessary once gender difference had been established on an essentialist basis. But the emphasis on the difference between the sexes continued to be emphasized in the innumerable discourses about gender published in the early modern period. However much these texts discuss women they are really written about men and intended for men. When women do appear, it is to show what men must not be and what they must not do. As Barbara Correll argues, women have "an essential function" in the conduct literature of the early modern period; they are "projected as the horror of effeminacy, which must be contained."[94] This comes out in Erasmus's handbook for the instruction of young boys, when he advises young males to model their behavior in opposition to that of girls:

> Attention must be paid to the care of the teeth, but to whiten them with fine powder is for girls. . . . It is boorish to go about with one's hair uncombed: it should be neat but not as elaborate as a girl's coiffure. . . . The hair should neither cover the brow nor flow down over the shoulders. To be constantly tossing the hair with a flick of the head is for frolicsome horses. The gait should be neither mincing nor headlong, the former being a sign of effeminacy, the latter of rage.[95]

Correll describes "a kind of psychopolitical crisis of masculine identity and authority" characterizing the early modern period.[96] The attempt to resolve it through increasingly rigid gender stereotyping was so successful that one of the female characters in Sigmund von Birken's "The Excellence of Women" (1669) complained that for women home is a virtual prison:

> Like inmates in a penal institution, we are taught handiwork, needlework, and spinning. We are banished to the kitchen and household chores, forced to become Marthas rather than Marys. How can we become virtuous when we are prevented from reading virtuous books? . . . And how are we supposed to talk intelligently, when we are forbidden to learn?[97]

Given the accumulated evidence, it seems fair to say that the emergence of new gender stereotypes based on an unprecedented dichotomy between the sexes was one aspect of the attempt to restore order and certainty in a world racked by skepticism, confusion, inconsistency, and hypocrisy. It was an imposed dichotomy that contradicted the reality of gender relationships. Mary Douglas makes the pertinent point that where male dominance is absolute, there is no insecurity and little or no fear of sexual pollution:

> When male dominance is accepted as a central principle of social organization and applied without inhibition and with full rights of physical coercion, beliefs in sex pollution are not likely to be highly developed. On the other hand, when the principle of male dominance is applied to the ordering of social life but is contradicted by other principles such as that of female independence, or the inherent right of women as the weaker sex to be more protected from violence than men, then sex pollution is likely to flourish.[98]

If we take sexual pollution in the widest possible sense as the fear of woman and female sexuality, we can see that this emerged with a vengeance in the witch trials of the early modern period and reflected male insecurity.

Although women were targeted as the great threat to the public order in the early modern period in reality men posed a much greater problem, and governments together with religious authorities made increasingly rigorous attempts to suppress the crime and disorder males, particularly young males, created. "Masculinity and its routine expressions," were according to Lyndal Roper, "a serious danger to civic peace rather than a prop of patriarchy."[99]

> At every turn . . . civic authorities found themselves confronted with the anarchic disruption caused by masculine culture—the feckless husband, the drunkard, the threatening collectivities of guild and gang. So far as its public manifestations were concerned, masculinity was far from functional for the patriarchal society of the sixteenth century.[100]

Peter Burke aptly describes the attempt to curb disorderly behavior as the "triumph of Lent" as the carnivalesque behavior associated with popular culture clashed with the campaign for the reformation of manners.[101] Carnivals, with their orgies of food, drink, dancing, and ribaldry, are characteristic of subsistence cultures, where resources are scarce and pleasure not an expected or possible part of everyday life. With the growth of urban culture and consumer society, however, the middle and upper classes had more resources at their disposal and began to distance themselves from popular culture. Medieval festivals of inversion, during which normal hierarchical relationships between males and females and rulers and the ruled were suspended, were outlawed, declared demonic, and equated with the witches' Sabbath. The turn toward the literal that marked the Protestant and Catholic Reformations along with the lack of humor generated by polarized thinking in general was a basic aspect of the campaign to reform manners. Nothing was fun or funny about theaters, alehouses, carnivals, or masquerades. To quote from one reformer's *Dissertation upon Drunkenesse* (1727):

> The vile obscene talk, noise, nonsense and ribaldry discourses together with the fumes of tobacco, belchings and other foul breaking of wind . . . are enough to make any rational creature among them almost ashamed of his being. But all this the rude rabble esteem the highest degree of happiness and run themselves into the greatest straits imaginable to attain it.[102]

Societies for the Reformation of Manners sprang up in England in the late seventeenth century with the aim of rooting out vice. These Societies employed paid staff to bring complaints and fill out warrants against drunkenness, swearing, gambling, breaking the Sabbath, prostitution, and any activity they deemed immoral. They organized raids of "Molly" houses, some of which led to trials and executions.[103] They also paid informers to collect incriminating evidence that could be used in court.[104]

The reform of behavior not only applied to the actions of individuals but also to the way they spoke. Renée Balibar and Dominique Laporte describe how the Académie Royale edited out excremental language from French vocabulary.[105] As social structure became more stratified with the growth of cities and the middle class, language became a marker of social difference. A gulf developed between the popular speech of the lower classes and educated speech.[106] Lawrence Stone claims that literacy actually decreased among the lower classes, making the distinction between social levels all the more obvious.[107] Across Europe language was sanitized, reaching a height of inanity in Thomas Bowdler's 1818 expurgated edition of William Shakespeare, which was deemed more suitable for women and children than what Shakespeare actually wrote.

Catholics were as eager to reform manners as Protestants, and they had even more institutions at their disposal to do just this. The most feared and dreaded of these institutions was the Inquisition, which was not in actuality a single organization but a number of different authorities located in different places and times. The Spanish and Portuguese inquisitions, for example, were most intent on ferreting out secret Moslems and Jews. This is not to deny that they were also concerned with magic and superstition as well as marriage and the family. We can see this in the breakdown of the 44,000 cases heard by the Spanish inquisition between 1540 and 1700. Of these, 2,645 (5.9%) were for bigamy; 1,131 (2.5%) for the solicitation of sexual favors by priests; and thousands more dealt with love magic. A significant number also concerned statements made by individuals that differed from the Church's teaching about sexuality. In both Aragon and Rome some 15 percent of the cases heard by the Inquisitions dealt with sexual matters. Ignorance about sexual matters was not considered a viable excuse because there was an annual reading of the "Edict of Faith" from every pulpit, during which Catholic teachings on a broad range of issues were described in simple language. People were therefore expected to know and conform to Catholic doctrine and practice. They were also encouraged to report on neighbors and acquaintances, who did neither. The various Inquisitions also conducted regular visitations and established a system of commissioners and unpaid agents known as "familiars." In some cities there was one familiar for every 50 people. Historians emphasize the climate of fear and suspicion created by this kind of surveillance. There simply was no need for draconian punishments, which explains why the various Catholic Inquisitions had lower rates of conviction and execution than secular courts. Just knowing that one was being watched was sufficient to keep most people in line. Overall, the Spanish Inquisition executed only 1.8 percent of those it tried, and it ceased executing witches in 1611, decades before secular courts, both Protestant and Catholic, did.[108]

The various Catholic Inquisitions were powerful tools in the attempt to civilize, reform, and educate Catholics, but they were not the only institutions to do this in the early modern period. The Jesuits were also instrumental in reinforcing patriarchal authority in families. They spearheaded a movement among young Catholic men to train them to become devoted upholders of Christian morality. Louis Chatellier claims that these young men became ubiquitous: "the seventeenth century was the century of the dévots. They were everywhere. . . ."[109] The dévots emerged from the Marian congregations established at the end of the sixteenth century with the goal "to reform the world."[110] These associations targeted males from all classes—merchants, artisans, nobles, and clerics—to inculcate pious habits and inspire missionary zeal. The dévots followed a daily routine of work and prayer, and they confessed and took communion frequently. Like many of their male Protestant counterparts, the dévots were obsessed with issues of impurity: "purity of self, purity in others; these were the aims."[111] They stressed chastity and modesty to the point of revealing a deep disgust for the body, and in their attempt to reform social morals, they attacked carnivals, masques, dances, swearing, taverns, prostitutes, and immodest female dress. The dévots shared other characteristics with evangelical Protestants. An essential element of their life involved the kind of work ethic historically associated with Protestants, one that emphasized hard work, the careful management of time, frugality, cleanliness, and godliness.[112] While the reformation of society was an important part of their mission, their principal mandate was to reform morality at home. They were, in short, instructed "in the wise and Christian direction of their families."[113]

Regulating families and family relationships was only one item on the agenda of those Catholics and Protestants set on reforming the manners of their fellow citizens. In this regard Isabel Hull is correct when she claims that the Reformation "created the first great codification of the rules of public order," and this applied to both Catholic and Protestant areas.[114] The effect was a crackdown on any kind of behavior deemed deviant and immoral that left little place for individuals who defied the increasingly stringent rules governing acceptable conduct. Only this change of perspective was needed to transform the popular medieval festival of misrule into the Witches' Sabbath.[115] As Sigrid Brauner and others have argued, the witch, as we know her, was invented in the early modern period as the antithesis of the ideal wife, later embodied in the eighteenth-century "Cult of True Womanhood."[116] Unlike the witch, who aggressively asserted herself and left home to visit and gossip, the ideal wife obediently stayed put, submissively ministering to the needs of her husband and children. She is the "Nouvelle Heloise," who would never make the mistakes of her predecessor, the old Heloise who succumbed to Abelard, bore a child, and died unrepentant. Thus, in both Catholic and Protestant families patriarchal authority was strengthened, and women were expected to obey their husbands and limit their activities to household tasks at the same time that both sexes were expected to act with circumspection and decorum.

A new world order had indeed been created.

CHAPTER 6

The Anthropological Revolution

In his book *Death and the Enlightenment*, John McManners describes the practice of magistrates in certain French towns of hiring men to patrol the streets at night. Dressed in black clothes with a white cross on the front and back or a tabard with a skull and crossbones, their job was to cry out at regular intervals, "Priez pour les trépassés" ("Pray for Sinners") and "Pensez à la mort" ("Think about Death").[1] Being reminded of death was not enough for some people, however. Actively thinking about it and actually going through the motions of dying, at least in one's imagination, were required as well. "It is good to die from time to time while you are alive," advised the Jesuit Jacques Noüet.[2] Another Jesuit, François Nepveu, recommended that old people single out the last day of every month for special prayers, acts of repentance, and meditation. They were to go to bed as if entering their coffin or tomb.[3]

Although this kind of fatalistic approach to life continued to characterize the attitudes of many people in the early modern period, in the late seventeenth and eighteenth centuries a new mentality emerged, especially among the affluent, who looked to this life rather than the next for satisfaction and fulfillment. As McManners says, "[T]hey were wanting to live longer, and they were discovering the logic to insist on enjoying life. . . . "[4] In her study of medical practice in eighteenth-century Braunschweig-Wolfenbüttel, Mary Lindemann contends that it was not just the affluent who wanted to live longer, healthier, and happier lives. In this community of predominantly small farmers, people were not resigned to death. They sought out medical help, even for small children at a time when infant morality rates were still very high.[5] As she says, when it came to health, "the inhabitants of Braunschweig-Wolfenbüttel approximated us more than their two centuries' removal in times might suggest."[6]

These examples show that the pessimism characterizing Reformation and Counter-Reformation attitudes had begun to dissipate, not for everyone to be sure and especially not for the poor; but for many people the idea of progress was in the air. People began to look forward to an age of change and innovation, when human life would be transformed and people might live long and happy lives in a more pleasant, comfortable, and salubrious environment. This optimism reflected a profound transformation in the way people viewed themselves, the world, and God. Was human nature irredeemably fallen and depraved as a result of original sin, or were human beings born good and corrupted by bad habits and a venal society? Was this world a vale of tears and human beings bound to suffer, or was it a place of infinite possibilities for self-expression and pleasure? Was God a harsh and exacting judge or a kind and loving father? In all three domains, the human, the natural, and the divine, things began to look up as a more positive and benevolent view of man, nature, and God emerged. Over the course of the long eighteenth century (1650–1800) there emerges an increasing resistance to the idea of human beings are innately sinful and a growing consensus that we are social creatures, whose essential goodness and natural sympathy for our fellow humans can be cultivated or destroyed by the treatment we receive from family and society at large. Hans Erich Bödeker has justly described this radical reevaluation of human nature as an "anthropological revolution."[7]

A number of different factors converged to create this anthropological revolution. The Protestant elimination of Purgatory severed the close ties previously existing between the living and the dead, and this, together with the Protestant emphasis on an individual's "calling" in this world, encouraged a more positive attitude toward the physical world and human activity in it. A similar validation of human activity occurred among Catholics as well, although for different reasons. Stung by Luther's attack on indulgences, Catholic reformers worked hard to put to rest the notion that salvation could be bought, emphasizing instead a new kind of activist philanthropy directed at transforming the lives of those on the margins of society. Some thirty new religious orders and congregations were founded from the late sixteenth through the eighteenth century, many with female branches and nine devoted exclusively to women.[8] The idealization of poverty and the salvific value of being poor characteristic of the medieval church was rejected by Protestants and Catholics alike. Poverty became a social issue that had to be addressed, and the poor, especially poor children, were now considered legitimate objects of charity with the potential to become productive, moral citizens.[9]

The strongest obstacle in the way of a more positive view of human nature was the doctrine of original sin. It was precisely this doctrine that came under concerted attack in the seventeenth and eighteenth centuries. As Ernst Cassirer says in *The Philosophy of the Enlightenment*, "the concept of original sin is the common opponent against which all the trends of the philosophy of the Enlightenment join forces."[10] Paul Hazard singled out the generation living between 1680 and 1715 as crucial in laying the foundations of modern thought, but his time scheme should be moved back to include Descartes, whose famous maxim, "I think, therefore I am," was decisive in this reevaluation of human nature. Descartes established an individual's consciousness—not God, not tradition, not society—as the Archimedean point in the universe. Human beings did not enter this world as either good or bad. They entered it endowed with the faculty of reason, and that was what shaped them into virtuous or reprobate adults. Descartes's method, based on systematic skepticism, was predicated on his conviction that neither faith nor tradition could provide individuals with certainty about what to think or do; only reason could do this. God was still in the picture to ensure that reason, when used correctly, was not deceptive, but Descartes essentially untethered human beings from God, leaving them to think for themselves.

Locke's *Essay on Human Understanding* was contingent on Descartes's "cogito." The question became, how do human beings think, understand, and know? This was what Locke set out to investigate, and his conclusion continued along the radical trajectory set by Descartes: human beings do not come into the world equipped with any predetermined character, sinful or innocent, or any innate knowledge whatsoever. They are the products of their experiences. At birth their brains are empty slates that are gradually written upon as material impressions enter their minds through their senses. Humans are nothing more nor less than the sum of their sensations, and the manipulation and regulation of these sensations provide the means to create new and improved human beings. Nurture not nature was responsible for whatever character individuals have, a fact that put the spotlight on education. As Locke says in *Some Thoughts on Education*, "of all the men we meet with nine parts of ten are what they are, good or evil, useful or not, by their education."[11] Writing to his friend Edward Clarke (September 1, 1685), Locke advised him on how to raise his son, emphasizing that the creation of good habits creates good character.[12] Locke's sensualist psychology turned children into natural guinea pigs for the innumerable schemes that emerged during the Enlightenment dedicated to the education of the "perfect" man and the "helpmeet" who would perfectly suit his needs—Rousseau's Emile and Sophie being prime examples.[13]

Gottfried Wilhelm Leibniz was another of the many philosophers who believed that education was a key element in building character, and he too rejected original sin. As he writes in his *Theodicy*, "Among the dogmas of the disciples of St. Augustine, I cannot swallow the damnation of infants . . . or in general damnation arising from original sin alone."[14] Gilles Deleuze takes Leibniz as the quintessential philosopher of the Baroque, because the notions of transformation or unfolding predominate in his work. In this regard, it is significant that the Latin *evolvere*, from which we get "evolution," means to unroll or unfold.[15] Unlike the inert, static atoms of the mechanical philosophers, Leibniz's building blocks of the universe, monads, are dynamic; they change, develop, and unfold as they strive to increase their level of awareness. Michael Gottlieb Hansch claimed that Leibniz even envisioned future human souls in a cup of caffe latte:

> I remember that once, when Leibniz and I met in Leipzig and were drinking caffe latte, a beverage which he greatly savored, he said that in the cup from which he was drinking there might be, for all we know, monads that in future time would become human souls.[16]

A dynamic view of human nature lies behind the German notion of "Bildung," or unfolding, a prominent theme in the classic German Bildungsroman that describes the development of the hero's character when faced with life's vicissitudes. A similar interest in the growth and development of character appears in novels in other parts of Europe as well. The novel had at least some of its roots in the spiritual autobiographies written by Protestants to gauge their spiritual progress. The same kind of scrutiny of character and motives one finds in Bunyan's *Pilgrim's Progress* is translated into secular and psychological terms when the innocent hero or heroine must make his or her way in the treacherous world. In a postscript to *Clarissa*, Samuel Richardson claimed that the novel was more effective than the pulpit in inculcating morality for just this reason. Richardson went so far as to say that he offered readers "the great doctrines of Christianity under the fashionable guise of an amusement." It is therefore understandable that Clarissa should liken her own suffering to Christ's. It was, she said, "in humble imitation of the sublimest Exemplar; I often say: Lord, it is thy will, and it shall be mine . . . I know thou will not afflict me beyond what I can bear."[17] Instead of seeing this as sacrilege—because surely Jesus had more on his mind than chastity—Clarissa's readers swooned in sympathy. Diderot, for example, wrote an "Eloge de Richardson" (1761) in which he reveals how deeply many eighteenth-century readers empathized with Richardson's characters to the point of modeling their own behavior on them. In previous centuries such a strong identification between author and reader had been considered a distraction from the real source of virtue, God. But this was not a problem for Diderot and his contemporaries who saw the source of virtue in other human beings. As Diderot said: "I have often said, as I read him: I would happily give my life to be like this woman; I would rather be dead than be that man."[18] The identification with fictional characters was so strong in the case of Jean Jacques Rousseau's readers that they literally wept and even had to take to their beds when Julie, the saintly protagonist in *La Nouvelle Heloise*, died. As Robert Darnton demonstrates, Rousseau's readers read him "religiously" precisely because he "demanded to be read as if he were a prophet of divine truth."[19] Rousseau was indeed sainted, and his grave at Ermenonville became a shrine.[20]

In these novels revelation is taken away from God and put in the human sphere as psychological introspection replaces Christian self-examination. The individual moves to center stage and bearing one's unique soul reveals the emotions that are the wellsprings of morality and not impediments to it, as they had been in earlier centuries, particularly among Protestants, who distrusted the emotions and the senses. The novel is thus *the* quintessentially modern text, fitting comfortably alongside economic theories of possessive individualism, private property, and political ideas of enlightened self-interest and individual rights.[21] Morality, as well as manners, had burst the boundaries of theology and philosophy to become a suitable field for novelists and essayists. Addison was pleased to make this point in relation to his essays:

> It was said of Socrates that he brought Philosophy down from Heaven to inhabit among Men; and I shall be ambitious to have it said of me, that I have brought Philosophy out of Closets and Libraries, Schools and Colleges, to dwell in Clubs and Assemblies, at Tea-Tables and in Coffee Houses.[22]

Addison's goal was to make philosophy "agreeable" to the burgeoning middle classes.[23] How far things have come from Luther and Calvin.

The idea that humans were not innately sinful but bound to improve under the proper tutelage became axiomatic for many Enlightenment thinkers intoxicated by the Promethean vision that, indeed, man makes himself. This represents a stunning change in attitude. To take the example of England: In *Paradise Lost* (1667) Milton had described man as sinful and fallen. Some seventy years later in his *Essay on Man* (1734) Pope had seen man as suspended between angels and beasts. By the end of the century many individuals, not only in England but across the continent as well, agreed with Erasmus Darwin that "All nature exists in a state of perpetual improvement," and this included human beings.[24] In his "Second discourse, sur les progress successifs de l'esprit humain" (1750), Turgot sketched out the history of human progress. Condorcet remained committed to his belief in the inevitable improvement of the human race even after he had been imprisoned during the Terror, eventually committing suicide to avoid the guillotine. In his *Essai sur la manière de perfectionner l'espèce humaine* (1756), Charles Augustin Vandermonde outlined the simple rules and natural principles that would make health, beauty, and strength hereditary, while simultaneously instructing parents how to train their children's minds.[25] The idea that the human race could be perfected appears in Germany as well and explains the particular fascination with reincarnation on the part of a broad range of thinkers, among whom were Wolff, Lessing, Schade, Clavius, and Weishaupt (to name just a few). Martin Mulsow argues that reincarnation provided an alternative moral framework to Christian sanctions against immorality, especially as the notion of Hell and eternal punishment lost credibility. For many thinkers the belief that people would reap their rewards or punishments in future lives offered a morally compelling and attractive picture of a God whose creation was preordained to return to its original perfection.[26] These optimistic assessments of human nature found an especially welcome home in the development of a philosophy of "Sentimental Ethics," which provided the foundation for "The Cult of Sensibility." In the eighteenth century "The Cult of Sensibility" emerged as a religion in its own right with the family an object of devotion, if not outright worship.

Anthony Ashley Cooper (1671–1713), third Earl of Shaftesbury, who had been a pupil of Locke, has been described as the "Father of Sentimental Ethics," which developed via Francis Hutcheson, David Hume, and Adam Smith into the school of Scottish "moral sense" philosophy. Shaftesbury took on himself the goal of vindicating both humans and God against their detractors, whether they came in the form of Hobbesians or Calvinists. In Shaftesbury's view, sympathy was a natural emotion; in fact, it was the quality that bound the world into a harmonious whole. Man was a social animal, and sympathy and "good humor" were the keys to a benevolent, well-regulated society.[27] The family was the arena in which sentimental ethics were most visible, but in Adam Smith's *Wealth of Nations* one can see how the sentiment of sympathy plays out in the economic sphere as well, leading to general prosperity. In the modern world, where economics seems to have nothing to do with ethics, the following passage about the family from *The Wealth of Nations* seems strangely out of place, but it demonstrates Smith's conviction that sympathy was the mark of a refined and civilized society. In this connection it is important to remember that Smith was Professor of Moral Philosophy at Glasgow University. The passage marks a profound change in religious sensibilities because Smith makes no mention whatsoever of religion or revelation. The sympathy he describes is an innate and natural human emotion:

> With what pleasure do we look upon a family, through the whole of which reign mutual love and esteem, where the parents and children are companions for one another, without any other difference than what is made by respectful affection on the one side, and kind indulgence on the other. . . .[28]

The most perfect figure in this family tableau is "the man of perfect virtue . . . he who joins to the most perfect command of his own original and selfish feelings, the most exquisite sensibility to the original and sympathetic feelings of others."[29] Here we can see at work what has been described as the secular religion of sensibility, in which human emotions are endowed with moral value. The "Cult of Sensibility" was an international phenomenon. It appeared throughout eighteenth-century Europe and its colonies in somewhat different forms, depending on the context in which the cult developed.

Shaftesbury was as concerned with defending God as man. He attacked both Hobbes and Calvin for creating a malevolent, demonic deity. For how could a benevolent God create innately evil men and then punish them eternally? The answer, as we have seen, is he didn't. Shaftesbury replaced the God of rewards and punishments—who was in his opinion a monster—with a "Divine Example," a kind of Platonic idea embodying truth and beauty. Replacing religion with aesthetics was one of the most significant features of eighteenth-century English culture.[30] This aestheticizing of the divine created a much sunnier view of religion than that prevalent during the Reformation. This comes out clearly in Addison's essay on popular superstitions:

> I know but one way of Fortifying my soul against these gloomy presages and terrors of mind; and that is, by securing to myself the friendship and protection of that Being who disposes of events and governs futurity. He sees, at one view, the whole thread of my existence, not only that part of it which I have already passed through, but that which runs forward into all the depths of eternity. When I lay me down to sleep, I recommend myself to his care; when I awake I give myself up to His direction. A midst all the evils that threaten me, I will look up to Him for help; and question not but He will either avert them, or turn them to my advantage.[31]

Addison was not weighed down by a sense of sin and fear of damnation as Luther and Calvin had been. He claimed that the immortality of the soul was a "subject upon which I always meditate with great Delight," something one could not imagine either Protestant reformer doing.[32]

The idea that God was a benign source of sympathy radiating through the universe colored ideas about what actually went on in heaven, a topic of consuming interest in the early modern period. Readers sent in questions to the *Athenian Gazette* on the subject, asking among other things if there were sexes in heaven and what were the effects of diseases and deformities and the mutilation of corpses on the battlefield when it came to resurrected bodies. The eighteenth-century emphasis on sentiment and the positive evaluation of emotions carried over into heaven. Instead of the eternal contemplation of the effulgence of divine light that Dante saw as the reward awaiting the resurrected, more active and domestic models of heavenly bliss emerged, emphasizing the continuation of the love between families, friends, and neighbors. There was even room in heaven for improvement, and a number of people began to envision heaven as a place where the natural philosophers would be able to continue experimenting. But James Boswell, a man known for both his religious scruples and his love of the flesh, had less cerebral pleasures in mind. He regaled Lord Kames with a description of a discussion on the subject of heavenly pursuits:

> I told him how Maclaurin had pushed Sir John Pringle at Lord Monboddo's upon the subject, and had asked him what we were to have that could make us wish for a future state: "Shall we have claret, Sir John?" "I don't know but you may, Mr. Mclaurin." "Well," said my Lord, "it is true this body is put into the grave. But may we not have another film, another body, more refined? The ancients," said he, "all describe a future state as having enjoyments similar to what we have here. Let us lay aside the prejudices which we have been taught. Suppose we have other bodies. Why may we not have all the pleasures of which we are capable here? For instance, the pleasure of eating. Why not that, in a more delicate manner?" I mentioned, before he spoke of eating, our being told we are to have music. "And," said he . . . "and there is another pleasure"; (I thought, though I divined what he meant clearly enough, that he should speak it out plainly, so waited in silence till he proceeded) "why not have the pleasure of women?" "Why not," cried I, with animation. "There is nothing in reason or revelation against our having all enjoyments sensual and intellectual."[33]

Boswell's hero Dr. Johnson was not able to take such a sanguine view of either God or the afterlife. Being reminded by his friend Dr. Adams that God was infinitely good, Johnson replied that did not mean he would not damn individuals, adding that he was afraid "I may be one of those who shall be damned." When Dr. Adams asked what he meant by damned, Johnson replied "passionately and loudly," according to Boswell, "Sent to Hell Sir, punished everlastingly." Jonson was incredulous when he learned that Dr. Adams did not share his belief in eternal damnation.

And when he said as much, only to have Mrs. Adams tell him that he seemed to have forgotten "the merits of our Redeemer," he thundered back, "Madam, I do not forget the merits of my Redeemer; but my Redeemer has said that he will set some on his right hand and some on his left hand." Boswell described him as in a state of "gloomy agitation" after this exchange and tells us that he cut the conversation short: "I'll have no more opn't."

Johnson was clearly in the right theologically, but he was out of step with a great many of his contemporaries, who, like Dr. and Mrs. Adams and Addison, took a more optimistic view of God, human nature, and the world at large and consequently dismissed the idea of an eternal hell. In their minds, if hell existed at all, it was there to reform sinners before they went to heaven. The role of hell was "medicinal," to quote Anne Conway, to "cure" individuals of whatever ailed them and caused them to act immorally; for a loving, benevolent, and just God could not possibly inflict infinite punishments on finite individuals whom he had created out of love.[34] The idea of universal salvation, for which the second-century church father Origen had been anathematized, was revived in the seventeenth century with the anonymous publication of *A Letter of Resolution concerning Origen and the Chief of his Opinions* in 1661. Origen saw history as a long and gradual process of education in which evil would be eliminated and universal salvation attained—even Satan would be redeemed.[35] Thus what D. P. Walker has described as "The Decline of Hell" was an important factor in fostering a new image of a benevolent God and the aestheticizing of religion. All this clearly bypassed Samuel Johnson and conservative Christians, but it affected many of their more liberal and optimistic contemporaries.

One of the most significant consequences of the rejection of the doctrine of original sin was an emphasis on human potential and the ability of individuals to shape their own lives. Roy Porter sums up this new orientation: "In the eighteenth century . . . [t]he scripts of life ceased to be God's: they were of man's own choosing."[36] This was the basis on which the "Enterprise Culture" emerged, a culture in which individuals had an increasingly high degree of responsibility for themselves and their social positions. While many individuals continued to attribute their actions to outside powers, such as spirits or demons, these attributions were less and less persuasive in an increasingly mobile and urban society, where achievement was based to a far greater degree than ever before on merit rather than kinship. The new sources of wealth generated from trade and emerging industries gave individuals more opportunities to better their positions. What increasingly mattered was not who an individual was but what he did and appeared to be.[37] There were still obvious tensions between the idea of an autonomous self and the class and patronage system current in the early modern period, but new opportunities, institutions, and social practices in the form of new jobs and professions, the increasing reliance on contracts, the establishment of academies and learned societies, and even the patronage of libraries and attendance at reading groups encouraged the idea of personal responsibility, social mobility, and self-fashioning. As the great German jurist and political philosopher Baron Samuel von Pufendorf (1632–94) claimed, man is not a social animal but a sociable animal. Thus society is no longer the condition of man, as Grotius had claimed, but the creation of men and consequently amenable to change and improvement.[38]

In 1945 Hannah Arendt predicted that the problem of evil would be the fundamental problem of postwar intellectual life. Susan Neiman has pushed the date much farther back, however, arguing that for the past three centuries philosophy has been an extended meditation on the problem of evil (and not a debate between skepticism and science or skepticism and religion, as many people contend): "If Enlightenment is the courage to think for oneself [as Kant claimed], it's also the courage to assume responsibility for the world in which one is thrown. . . . Modern conceptions of evil were developed in the attempt to stop blaming God for the state of the world, and to take responsibility for it on our own."[39] Neiman places Rousseau at the center of this revolution. For Rousseau evil is still the result of the human abuse of free will. Sexuality, jealousy, and possessiveness are the causes of our suffering. But Rousseau vindicated God without damning men as earlier theodicies had. Evil is our own doing, but man is not innately evil. Knowledge is what is needed, not repentance. Evil must be treated as an historical, not religious, problem. Only then can human beings attempt to make things better.[40] Neiman recognizes that Rousseau was not alone, that, indeed, his view of evil and the possibility of eradicating it was a motive force among early modern natural philosophers.[41]

The belief that evil was an historical and human problem that could be eradicated through human effort encouraged many people across Europe to join newly established scientific organizations and to send reports to

England's Royal Society and other scientific institutions, describing experiments, inventions, and discoveries to enhance life through improvements in agriculture, manufacturing, and education, the reform of the medical profession, and the fostering of religious tolerance. Commenting on this fascination with science, Russell Shorto writes, "If some one from any of the previous centuries could have revisited earth in the 1700s, it might reasonably have seemed that human beings had become drunk on invention."[42] Two of the most indefatigable experimenters were Robert Hooke and John Wilkins, both members of the Royal Society. John Evelyn, another member of the Society, describes the many scientific schemes they pursued after they escaped from London during the plague year of 1665 and set up shop in Durdans, near Epson. He describes the two hard at work "contriving chariots, new riggs for shipps, a wheele for one to run races in, and other mechanical inventions." In addition to constructing these practical devices, they engaged in a program of

> divers experiments of heat and cold of gravity and levity, of condensation and rarefaction or pressure, or pendulous motions and motions of descent; of sound of respiration, or fire, and burning, of the rising of smoke, of the nature and constitution of the damp, both as to heat and cold, driness and moisture, density and rarity, and the like.[43]

The Philosophical Transactions of the Royal Society are full of experiments like these, as were books and periodicals from across the channel. Leibniz, for example, kept a diary in which he recorded collaborating with his friend Francis Mercury van Helmont on plans for better cooking pots, more efficient wheelbarrows, and even shoes with springs "for fast get-aways."[44] Like so many of his contemporaries Leibniz was committed to doing everything possible to improve the human condition, first and foremost by restoring religious unity and, second, by devising innumerable projects for all kinds of socially useful inventions. His calculator is perhaps the best known, but he also proposed plans for such things as a high speed coach, which would run along rutted tracks on something like ball bearings, a scheme for draining water from the Hartz mines, an inland navigation system, the manufacture of porcelain, the exploitation of waste heat in furnaces, tax reform, a public health and fire service, steam-powered fountains, street lighting, a state bank, and isolation wards for plague victims. Samuel Hartlib (1600–62), the central figure in a group of reformers that included the Czech teacher and educator John Amos Comenius (1592/4–1680) and the peripatetic Scottish ecumenical missionary John Dury (1596–1680), was also a facilitator for all kinds of improvement schemes involving agriculture, medicine, language, religious toleration, and education. Hartlib, Comenius, and Dury formed the nucleus of a European network of scholars and philosophers who worked for reform in all spheres of life. They were known as "Pansophists," taking their name from Comenius's lifelong endeavor to construct an outline of universal knowledge, or "Pansophia."

In *Gulliver's Travels* Jonathan Swift ridiculed efforts like these with his scathing descriptions of the academicians in The Grand Academy of Lagado fruitlessly trying to reconstitute food from excrement and extract sunbeams for cucumbers.[45] In his immensely popular play *The Virtuoso* (1676), which ran on and off for some twenty years, Thomas Shadwell created the character of Sir Nicholas Gimcrack, the prototype for the harebrained scientist with his misplaced enthusiasms. As Sir Nicholas proudly proclaims, " 'Tis below a Virtuoso, to trouble himself with Men and Manners. I study Insects." Indeed Sir Nicholas spends two thousand pounds of his niece's money to observe mites in cheese through a microscope, read the Bible by the light of a rotting leg of pork, and transfuse sheep's blood into a madman, all parodies of actual experiments conducted by members of the Royal Society. Shadwell modeled Sir Nicholas on Robert Hooke, who was so mortified when he went to the play that he wrote in his diary: "Damned dogs. Vinda me deus [God grant me revenge]. People almost pointed."[46] Although Charles II was very interested in science, he referred to the Fellows of the Royal Society as his "fous" or "crazies," which catches the tenor of some of the contributions to the *Philosophical Transactions*.[47] John Hill collected some gems in his book *A Review of the Works of the Royal Society*. A notable one describes "The Bacon Scheme of Duck Catching":

> The Method is this: tie to the end of a long String a Piece of fat Bacon, two inches long and half an Inch thick, let the other End of the String be fastened to a Tree or Post, then leave it the whole Night: the first Duck that comes will eat the Bacon, and, as the worthy member who proposed the method observes, will void it again in a few minutes; it will then be gobbled up by another, t'after that a third, and so on, each voiding it soon after the swallowing, and . . .

> regularly passing thro' the Guts of the whole Covey: thus in the Morning a whole string of Ducks will be found ready catched, and there needs only the drawing of the string to take them all up.[48]

Admittedly, examples like this are very funny, and, as in this case, it was easy to poke fun at the crackpot schemes proposed by natural philosophers. But Swift's satire along with the Shadwell and Hill's comic take on the more extreme examples of scientific enthusiasm provide evidence of the excitement and optimism inspired by science at the time. Francis Bacon's writings were immensely important in this regard with his admonition that, unlike their medieval predecessors, natural philosophers had the duty to produce practical "fruit" from their scientific researches as well as intellectual "light." Bacon practiced what he preached, becoming a martyr to the cause of science. Stuffing a dead chicken with snow to see if cold could arrest its decay, he caught the chill from which he died.

Bacon was a major figure in promoting the "modern" agenda that things could and would advance and improve; and this included the life span of ordinary individuals.[49] Although he felt the need to bow to the Christian view that death and salvation were the ultimate objectives of human life, he was convinced that while alive, individuals should have the means to keep their bodies and minds as healthy and active as possible.[50] In his *New Atlantis* Bacon lists among the "Magnalia Naturae," or those things men most desire, "the prolongation of life," "the restitution of youth," and "the retardation of age." He describes the scientists employed in "Salomon's House," the scientific institution in his imaginary "New Atlantis," who produce a "water of Paradise," that is "very sovereign for health, and the prolongation of life."[51] There were also special caves on the island with air that helped heal and prolong life and a bath that restored "the very juice and substance of the human body."[52] Salomon's House contained laboratories in which animals and birds were dissected "that thereby we may take light what may be wrought upon the body of man." The fact that headless birds "leape and flutter" suggested the possibility of resuscitating the dead. Bacon even proposed that organs could be transplanted.

Bacon's *History of Life and Death* did much to legitimize the idea that prolonging life was a legitimate and worthy goal.[53] He viewed illness and death as both natural and avoidable. This makes it all the more significant that he wrote at a time when innovation was still looked upon with suspicion and tradition, summed up in the wisdom of the ancients, provided the touchstone for the conservative ideology of the court society to which he belonged. Reverence for the past was, however, in the process of giving way—at least in the minds of some people—to a passionate embrace of the new and innovative. Bacon was a pivotal figure in this transition as he carefully weighed ancient traditions against new ideas and came to the conclusion that older and earlier were not always better.[54] Thus, although his ideas were admittedly highly derivative, going back to medieval and Renaissance sources, they resonated in subsequent decades in the minds of those who championed the Moderns and repudiated the Ancients.

Like Bacon, Descartes embraced the idea of progress and believed people could live longer, healthier, and more productive lives. His optimistic, progressive agenda appears in a letter to Mersenne in March 1636, where he describes what came to be known as his *Discourse on the Method of rightly conducting one's reason and seeking the truth in the sciences*. This eventually became one of four treatises to be included under the general title "The Plan of a universal Science which is capable of raising our nature to its highest degree of perfection."[55] Descartes was convinced that "we might free ourselves from innumerable diseases, both of the body and of the mind" to the point that we might eventually be freed from "the infirmity of old age."[56] Charles de St. Evremond (1610–1703) described a visit that Sir Kenelm Digby (1603–65) made to Descartes, in which the prolongation of life was discussed. Digby initiated the subject by suggesting it was more important to try to prolong life than to speculate about abstruse philosophical issues. St. Evremond reported that "M. Des Cartes assured him [Digby], that he had already considered that matter; and that to render a man immortal, was what he would not venture to promise, but that he was very sure it was possible to lengthen out his life to the period of the Patriarchs."[57]

Bacon and Descartes were key advocates of the idea that nature could be manipulated and controlled, a guiding principle of the Scientific Revolution. Their optimism flew in the face of the skeptical conditioning that was part and parcel of a classical education and Christian upbringing, both of which emphasized the precarious nature of life on earth and the helplessness of individuals in the face of suffering and pain unless they possessed philosophical resignation or a religious cast of mind. But the myth of a lost Golden Age and the idea that the Ancients were vastly superior to the Moderns, which had held sway for so long, came under increasing attack in the early modern period as trade, travel, developing commerce, and the ingenuity of artisans, craftsmen, and natural philosophers opened up vistas of unimagined possibilities and brought new and unprecedented things such as rhinoceri and anteaters, not to mention potatoes and tobacco, to the attention of Europeans.[58] Microscopes and telescopes may have blown the minds of natural philosophers, allowing them to see things they never knew existed, but air and water pumps, carriages with better springs for faster and more comfortable travel, newly designed houses with smaller, heated rooms, and more sophisticated, healthful, and appetizing food convinced many ordinary people that, indeed, life was worth living and that science was a potent force for good.[59] In addition to these practical improvements, the magisterial work of Sir Isaac Newton in physics and optics was heralded as irrefutable proof that the modern world was unquestionably many steps ahead of the ancient. Bacon's famous comparison between the ancients and the moderns who stood on their shoulders and could therefore see farther was picked up by many of his contemporaries.[60] In this respect it is interesting to compare Petrarch's reverence for the ancients with Bacon's rather offhand dismissal of them. In his letter to Homer, Petrarch describes himself as a "babe," who loves to "babble with those who feed men, even though they are skilled masters of speech."[61] Bacon, on the other hand, describes the wisdom of the Greeks as "like the boyhood of knowledge" with "the characteristic property of boys: it can talk, but it cannot generate, for it is fruitful of controversies but barren of works."[62]

"Improve" and "Improvement" became watchwords of the eighteenth century, and the idea of "deliverance" or "escape" from the ignorant traditions of the past a common theme. Locke describes himself as an "Under-Labourer clearing Ground a little, and removing some of the Rubbish that lies in the way of knowledge."[63] Bacon talked about the "Idols" standing in the way of progress, one of the most important being "The Idol of the Theater," those false ideas implanted in people by false philosophy, theology, and superstition. Old wives' tales and folk wisdom fall into this category, and Bacon, like many of his contemporaries, was critical of both. Proverbs, revered up to the middle of the seventeenth century as a source of collective wisdom, were rejected in favor of the independent judgment of the individual. As Lord Chesterfield said, "a man of fashion never has recourse to proverbs or vulgar aphorisms." Doing so would reveal him as stultifying unoriginal and lower class to boot.[64] What became increasingly valued in the early modern period was what was new and original. The connection was constantly made between the New World and the new discoveries made possible by the New Science. Nicholas Monardes, for example, gave an enthusiastic endorsement of the "newe thyges and secrets" discovered in the Americas in his *Joyfull Newes out of the Newe Fonde World* (1577). He was especially taken by "the rare and singular vertues of diverse and sundried Hearbes, Trees, Oyles, Plantes, and Stones, with their applications as well for Physic as Chirurgerie."[65] One of the reasons why cabinets of curiosities were so popular was to highlight the juxtaposition of classical fragments with modern instruments, such as telescopes, globes, lenses, microscopes, and specimens of flora and fauna from the new world, emphasizing how the once dominant culture of the past was receding before new forms of modernity.[66] Jonathan Israel draws attention to the essential role of journals in spreading an awareness of new ideas and turning people away from old authorities.[67] While it is true that all these developments only affected a small proportion of the population, that minority directed public opinion, and public opinion was definitely tipping toward the view that the modern world held distinct advantages in making the good life here and now a real possibility.

As we have seen, Bacon thought revitalizing the body's "juices" might be one way to improve health and prolong life. After William Harvey's discovery of the circulation of the blood, this idea was taken up in the many experiments involving blood transfusions. Transfusing young blood into aged veins produced some extraordinary results. In the 1650s the English physician Richard Lower was among the first to successfully transfuse blood between dogs. Descriptions of such transfusions were published in *Philosophical Transactions* of the Royal Society, exciting much interest. In one account, "an old mongrel curr, all over-run with the mainge" was "perfectly cured" by the blood of a young spaniel. The same success followed another experiment, in which "Mr. Gayant transfused the blood of a young dog into the veins of an old, which two hours after, did leap and frisk; whereas he was almost blind with age, and could hardly stir before."[68] It wasn't long before blood transfusions were tried out on human beings, but without any knowledge of the importance of blood types, it was only a matter of time before someone died, which led to a moratorium on further experiments that lasted until the early nineteenth century.

The great hope invested in transfusions may have been extinguished by such disasters, but Harvey's discovery of the circulation of the blood (on which the possibility of transfusion hinged) was part of a major shift in the way the body was viewed in the early modern period. Harvey was not the first person to have dissected a heart. The second-century physician Galen had done the same thing. But Harvey lived in an age of air and water pumps and consequently had a mechanical model he could apply to the working of the human body. The ability to envision the heart as pump had profound implications: it meant that not only could the body, like every other mechanical device, be understood and investigated, but it could be fixed and even improved when problems arose.[69] Instead of being the corruptible locus of sin, sex, putrefaction, and disgust, the body was now viewed as a machine. Machines are not by nature disgusting; machines are not associated with drooling, defecating, farting, burping, incontinence, and uncontrollable libidinous urges. They run according to a predetermined plan; and while they may become disabled and cease to function properly, there is always the hope that they can be patched up or new parts substituted for worn ones. The fear and loathing of the body characteristic of so much classical and Christian thought diminished in the climate of this medical materialism, and a new pride in the body arose. Robert Boyle, for example, thought it was dishonorable for "a Reasonable Soul to live in so Divinely built a Mansion, as the Body she resides in, altogether unacquainted with the exquisite structure of it."[70] Addison declared the human body "Handsome, harmonious, durable, adaptable to all needs and capable of self repair, man's flesh and blood was itself a hymn of praise to the Great Original."[71]

The English physician Thomas Willis recognized that Harvey's discovery required a complete rethinking of human and animal physiology. He put the word "neurologie" on the scientific map in such books as *Cerebri Anatome* (1664) and *Pathologiae Cerebri et Nervosi Generis Specimen* (1667), in which, among other things, he proposed that epilepsy and other convulsive disorders had neurological, not supernatural, origins.[72] Porter emphasizes the importance of these developments: "by the close of the seventeenth century, advances in anatomy and physiology had created the promise of a scientific understanding of the body, matching what high-prestige mathematical astrophysics and mechanics had done for the inanimate world."[73] The advances in medicine and conviction that health could be improved was an aspect of what Peter Gay has described as the "medicalization" of the Enlightenment by those authors who discussed the shortcomings of society and social institutions—religious fanaticism, political injustice, poverty, the prevalence of ignorance and superstition—as "pathologies" that could be "cured."[74] The most effective way to cure these pathologies—and as we have seen sin was now considered one of these—was by improving human reason and perfecting human nature.

The growing faith in science was fostered by nothing so much as the success of small pox inoculation. Small pox was a dreaded disease. According to d'Alembert it affected as many as eight out of ten people, killing one out of seven of those afflicted.[75] The scarring associated with the disease figured in the plotlines of novels and dramas of the period. In *Les Liasons dangereuses* the worst thing one could wish on one's enemy was to get small pox and recover with "un visage à faire tremblé" ("with a face to make everyone tremble"). In Prévost's *Mémoirs pour server à l'histoire de Malte*, the hero returns to find his love, Helena, completely disfigured by the disease; he can only make love to her in the dark. Inoculation was one of the great medical success stories of the eighteenth century and did much to encourage the belief in progress. Abbé Roman chose this as the subject of his celebratory poem, 'L'Inoculation, Poème en quatre chants." A new hairstyle for women was devised to commemorate the discovery, "coiffure à l'inoculation," which consisted of an olive tree with a serpent entwined about its trunk—symbols of wisdom and Asclepius, the God of medicine and healing in ancient Greece—with the sun, signifying enlightenment, rising in the background.[76]

A number of scholars have discussed what they describe as the "domestication" of science during the eighteenth century as air pumps, barometers, and electrical machines were brought into households for experiments and even incorporated in furniture. So great was the interest in new scientific apparatus that instrument makers increasingly catered to an amateur and domestic market.[77] A thriving trade in scientific books emerged as authors began writing for ordinary people, expressing scientific concepts in familiar, accessible language. John Heilbron attributes the vogue for chubby little boys, or *putti*, in scientific illustrations to this desire to domesticate science and make it appear both innocent and licit. He offers a marvelous selection of images showing *putti* performing various experiments, including one in which they use an air pump to suffocate a small animal, unfortunately a popular practice at the time.

The inclusion of *putti* made such experiments more palatable.[78] The popularity of scientific books was so great that Buffon's forty-four volume *Histoire naturelle* was reprinted several times during the century. The naturalistic illustrations of animals against landscape backgrounds in Buffon were in such demand that they were published in volume six of the *Encyclopédie* (1768) as well as separately.[79] In addition to the domestic consumption of scientific instruments and illustrations, there was great interest in scientific experiments performed at the increasing number of observatories, museums, and botanical gardens open to the public.[80] The importance of private citizens and especially of women in domesticating science is emphasized by Margaret Jacob and Dorothée Sturkenboom, who argue that in judging women's place in science more attention should be paid to the scientific activities of women at home than to the societies and academies that excluded them.[81] On the basis of their research, they claim that in terms of "the distribution and absorption of scientific knowledge," the women in the Dutch town they studied were "at the center of a major intellectual transformation."[82] They were involved in the "domestication of science," a vital process "for the deep embedding of science in Western culture."[83] In this truly revisionist history, women play a central role in the dissemination of the very institution that defines the modern Western world:

> Our histories take it for granted that sometime after 1750 Westerners of all ages became interested in science, but we do not see that households, leisure-time activities, conversations at dinner tables, and advances in female education may have been critical elements in that slow but remarkable process, apparent by the late eighteenth century throughout the old and new worlds. Before 1800, when science in the form of physics, mechanics, and advanced mathematics had not become part of every secondary school curriculum, it resided in other venues. Its mores, ideology, and practice had seeped into elite households—consequences that can still be traced in educated culture.

Science came into its own in the early modern period, and although a number of historians reject the idea that the seventeenth century marked the age of "The Scientific Revolution," or that such a revolution even occurred, the event has more basis in reality than its detractors recognize. While the term "revolution" may be misleading because there was no single event to mark the Scientific Revolution or specific time period (since different branches of sciences developed at different paces), the term is still useful to indicate that a major transformation occurred in the way people thought about science, themselves, and the natural world. During the long eighteenth century science became a part of the general culture in a way it never had been before; as we have seen, it became "domesticated." For some scholars the idea of "domestication" is far too benign to describe what actually happened, namely, the advent of "Scientism," or the domination of science over all other fields of human endeavor. In their view this domination has proven disastrous because it laid the foundation for the totalitarianism and genocide blighting the modern world. According to this scenario, science has become a religion in its own right, and a pernicious one at that, with scientists assuming the mantel of monks, Galileo a saint and martyr, and Newton the new Christ.[84] These are issues that will be discussed in the final chapter, but what is important at this stage is to understand the role religion played in the emergence of science.

This chapter has discussed the way "the anthropological revolution," with its rejection of the doctrine of original sin, encouraged individuals to embrace science and accept notions of progress and change. The next chapter will challenge the commonly held belief that religion and science are constitutional enemies continually at loggerheads and describe the role of religious beliefs in fostering the Scientific Revolution. While certain aspects of Christian theology provided a general impetus for the pursuit of science, what really provided a catalyst for the enthusiastic acceptance of science in the seventeenth century was apocalyptical thinking and millenarianism. Today apocalypticism is associated with terror and violence, but in the seventeenth century it was a crucial factor not only in the emergence of modern science but in promoting liberal values and political democracy. A further factor contributing to the development of modern science must be taken into account as well, and that is the occult and esoteric sources that fostered the idea of human agency and the possibility of change and progress. The old idea that religion and magic, as well as esoteric thought of all kinds, had to disappear before science could emerge is quite simply wrong.

CHAPTER 7

Religion and the Scientific Revolution

The previous chapter described the anthropological revolution that occurred in early modern Europe and argued that one of the major factors behind this revolution was the optimism and confidence in human ability fostered by developments in science. This conclusion, which appears obvious to many people, is not universally accepted and has in fact come under fire in recent years. The very concept of a "Scientific Revolution" has been rejected by a number of scholars, who question the use of the term "revolution" to describe a series of events that took anywhere from 150 years, the time between Copernicus and Newton, to 500 years if one wants to included the "delayed" revolution in chemistry and biology or if one considers medieval science the *sine qua non* of modern science as did Pierre Duhem.[1] For all that has been written on this issue, those who support the idea that the Scientific Revolution occurred and that for better or worse it has shaped our modern mentality and the modern world carry the day, at least in the opinion of this author. Historians may have lost some of the enthusiasm for science exhibited by Herbert Butterfield when he claimed that the significance of the Scientific Revolution outdid "everything since the rise of Christianity" and made the Renaissance and the Reformation shrink "to the rank of mere episodes."[2] But, as Richard Westfall argues in his defense of the term, a real transformation occurred as a result of developments in science, and this transformation was "a once and for all event that has never been reversed." Westfall makes the telling observation that "Scientists of today can read and recognize works done after 1687. It takes a historian to comprehend those written before 1543."[3] He itemizes key aspects of this transformation: the development of a new science of mechanics; the first application of mathematical laws to motion; the recognition that the same forces apply to particles as to planets and can be expressed mathematically; the emphasis on experiment and the international character of experimentation; the improvement and invention of instruments such as the telescope, microscope, barometer, air pump, and instruments for surveying and surgery; the blossoming of technology; and, finally, a new relation between science and religion with science asserting its autonomy. Westfall concludes, "The whole meaning of the Enlightenment of the eighteenth century was the authority of science over the intellectual life of Europe."[4] John Henry compresses Westfall's list of the most salient aspects of the Scientific Revolution into three main ones: (1) the increased use of mathematics to explain real effects in the natural world; (2) the new emphasis on observation and experience rather than tradition and authority; and (3) the idea that knowledge of the world should be put to use.[5] Peter Dear takes a more sociological approach and sees the Scientific Revolution as a radical transformation in the type of knowledge that was valued: from the goal of understanding the world and establishing truth, natural philosophers turned to the task of investigating how things worked with the goal of controlling them.[6] Margaret Jacob is another historian who supports the idea that something transformative happened, putting this in the broader psychological terms sketched out in the previous chapter: "More than any other body of culture, science released the revolutionary imagination, helped to develop its fantasies, to eliminate doubt about what human beings could accomplish."[7] These various appraisals show that there are many ways to broach the Scientific Revolution, ways that run from the strictly scientific to the social, historical, and psychological, and in recent years most historians of science have become resolutely eclectic, using all these approaches.

The term Scientific Revolution first appeared among historians of science in the work of Alexandre Koyré in the 1930s and became popular after World War II.[8] It had been used in a positive sense two centuries earlier, however, by d'Alembert in his "Preliminary Discourse" to the *Encyclopédie*. Already by the mid-eighteenth century the narrative of the triumph of science and rationalism in the West had taken shape, and a new type of cultural hero emerged, the selfless, objective, stoical man of science, dedicated to the improvement of human life.[9] A number of scholars have ridden roughshod over these ideas, however, dismissing the first as an unwarranted avowal of Western exceptionalism and the second as a fiction invented to obscure the truth that scientists, like everyone else, are creatures of their culture and reflect cultural biases. As Jacob has pointed out, however, "Heroes are not born, they are made." The fact that Galileo, Kepler, and particularly Newton were lionized, if not deified, in the eighteenth century indicates that there had been a major transformation in the way literate Westerners looked at the world and history and in their choice of heroes to represent the best aspects of both.[10] In his defense of the legitimacy of the term H. Floris Cohen characterizes "the Scientific Revolution" as the crucial event in the transition from traditional to modern modes of thinking and acting:

> . . . the Scientific Revolution is the embodiment of humanity's beginning to gain a consolidated, coherent grasp of nature. . . . The question of how this breakthrough occurred, and how it could occur, is what has continued to occupy historians of science. . . . And it is the very special event of its emergence in the sixteenth and seventeenth centuries that has set the West apart—the decisive event that sealed for good Europe's already on-going drift away from traditional modes of acting and thinking. . . .[11]

Although Cohen is undoubtedly right in claiming the Scientific Revolution created a rupture with traditional modes of thought at least in the long term, it is equally true that the Scientific Revolution cannot be understood without taking into account the vital contributory role religion played in its development. During the past fifty years historians of science have increasingly dismissed the once dominant idea that modern science and modern ways of thinking emerged only when rationalist philosophers rejected religion and magic. We are now aware of the important way in which religion, magic, and science interacted to produce the Scientific Revolution. As John H. Brooke has argued so effectively, Christianity provided the presuppositions, justifications, and motivations for science by establishing that God had created an orderly universe, which was itself "a book of nature" that must be studied in conjunction with Scripture to reveal the majesty and goodness of God. Furthermore, the idea of history as linear and progressive, the emphasis on the role of man as nature's steward, and, most importantly, the belief that humans were created in the image of God all contributed to legitimizing and promoting scientific activity.[12] The pursuit of science was justified in the early modern period as "natural theology." Kepler considered himself a priest "of the Most High God with respect to the Book of Nature," claiming that by discovering cosmic laws he was "thinking God's thoughts after Him."[13] Robert Boyle extended this line of thought to all natural philosophers, who were in his view the real "priests" because they studied nature to reveal proof of God's goodness, benevolence, and the scope of divine providence.[14] John Ray titled his book on natural history *The Wisdom of God Manifested in the Works of Creation* (1691).[15] Paracelsus, van Helmont, and their followers advocated the empirical study of nature as the only way to learn about God first hand;[16] and Newton was convinced his scientific discoveries offered invincible proof of God's continual involvement with the created universe.[17] These were not pious platitudes but evidence of the deep religious commitments that motivated early modern natural philosophers. Malebranche claimed that "One insect is more in touch with Divine Wisdom than the whole of Greek and Roman history."[18] Jan Swammerdam waxed eloquent on the subject of a louse: "Herewith I offer you the omnipotent Finger of God in the anatomy of a louse: Wherein you will find Miracles heaped upon Miracles and will see the wisdom of God clearly manifested in a minute point."[19] Some of the examples adduced to prove how well designed the world is are amusing, even outrageous, by today's standards; but we have read Voltaire's *Candide* and enjoy the pleasure Voltaire took in ridiculing the arguments from design put in the mouth of Dr. Pangloss. Henry More sounds just like the good doctor when he claimed, some sixty years earlier, that it is a mark of divine providence that the forehead is "fenced off by those two wreaths of haire which we call eye-brows" so that sweat does not run into the eyes. Or take More's claim that "the eye-lides are fortified with little bristles as with Palisades, against the assault of Flyes and Gnats," or his contention that "Nature has made the hind-most parts of our body which wee sit upon most fleshy . . . making us a natural Cushion."[20] At least More did not claim, as Luther did, that fleshy hind parts were the distinctive feature of women and a divinely ordained sign that they should sit at home.

The design argument was not new in the early modern period. It was a feature of medieval theology as well, but the sheer exuberance with which it was presented in the seventeenth and eighteenth centuries is indicative of the enthusiasm with which natural philosophers went about their work. In *A Demonstration of the Being and Attributes of God*, Samuel Clarke shows how closely connected the optimistic reading of the book of nature was to a positive view of human abilities:

> . . . the notices that God has been pleased to give us of himself are so many and so obvious in the constitution, order, beauty and harmony of the several parts of the world, in the form and structure of our own bodies and the wonderful powers and faculties of our souls, in the unavoidable apprehensions of our own minds and the common consent of all other men, in everything within us and everything without us, that no man of the meanest capacity and greatest disadvantages whatsoever, with the slightest and most superficial observation of the works of God and the lowest and most obvious attendance to the reason of things, can be ignorant of Him; but he must be utterly without excuse."[21]

An even stronger motivation to pursue science than natural theology was the apocalypticism and millenarianism that were such prominent features of early modern thought. That apocalypticism—the conviction that the end of the world was imminent—could motivate people to study the natural world may seem an unlikely proposition today, when apocalyptic thinking is linked to a suspicion of science and focused on escaping this world. However, as Arthur Williamson shows in his insightful book on early modern apocalypticism, what is now considered a "creed for cranks" played a crucial role not only in the development of early modern science but in the emergence of liberal values and political democracy. Between 1500 and 1800 many of Europe's most innovative thinkers across the religious spectrum among both Christians and Jews believed they were living at the end of days, and this conviction spurred many of them to action. Williamson claims that apocalyptic thinking "underwrote the Reformation in the sixteenth century, the British Revolution in the seventeenth century, and the American Revolution in the eighteenth century." He goes even further, claiming that apocalyptic expectations provided the "central motor of modernity" and played a pivotal role in the emergence of secular culture.[22] The irony of course is that religion was instrumental in the growth of skepticism about religion as well as the rise of secularism.[23]

As a result of the wave of apocalyptic thinking that spread across Europe, there was an outpouring of "spiritualist-scientific utopias" predicated on restoring humans to their prelapsarian state in the Garden of

Eden. Many scholars have discussed utopian and millenarian schemes on the continent among both Christians and Jews. Richard Popkin spent the last thirty years of his life documenting the interaction between Christian millenarianism and Jewish Messianism, particularly in Holland and England. While antisemitism was a fact of life in early modern Europe, Jews and Christians managed to interact in ways that were intellectually and culturally fruitful for both communities, and millenarianism was one such area where the cross-pollination of ideas took place.[24] Catholic millenarianism lies behind the emergence of the great Spanish and Portuguese colonial empires, in which the ideas of dominion and salvation were inextricably entwined. By 1530 Spain had conquered the New World, consolidated its power in central Europe, Italy, and the Netherlands, and turned back the Turks at Vienna. Such a display of power encouraged eschatological expectations. For what other reason, Popkin speculated, were the monks in the Alhambra learning to speak Hebrew than to talk directly to Jesus upon his return to earth? Apocalyptic expectations help to explain Louis XIV's war on the Dutch Republic and the Revocation of the Edict of Nantes, both actions in which it is impossible to separate religious and political motives.

While there is nothing scientific, liberal, or democratic about the apocalypse itself, apocalyptic expectations fueled visions of a glorious future time when humans would work together cooperatively without distinctions of rank or class. The most distinctive aspect of early modern millenarianism and what sets it apart from earlier millenarian thinking is this kind of activism. Seventeenth-century millenarians, whether Christian or Jewish, did not sit back complacently and wait for the millennium to come to them; they sought to make it happen. The ideas of "restoration," "restitution," and "renewal" were key in this vision and emerged from both the Old and New Testaments. *Tikkun olam* is a Hebrew phrase meaning "repairing" or "restoring the world." The concept emerged in early rabbinic Judaism and was given new significance in the mystical writings of Jewish Kabbalists, whose works were translated and disseminated throughout Europe in the early modern period.[25] The role ascribed to human beings in the Kabbalah and particularly in the Lurianic Kabbalah of the sixteenth century is startlingly positive.[26] The complex mythology of the Lurianic Kabbalah is predicated on the notion that the very act of creation involves separation, division, and a consequent Fall. The Fall is thus an inherent aspect of creation that occurs *before* man is created, and human nature is therefore not implicated and consequently not incapacitated. On the contrary, human beings are assigned the role of restoring the world to its prelapsarian perfection through the process of *tikkun*. The central role given to man in this process obviates the need for an external savior and encouraged human beings to act on their own initiative rather than wait for divine intervention.[27] The Kabbalah's emphasis on human responsibility for the cosmic order leads Moshe Idel to describe man's role as "universe maintenance activity."[28] Zwi Werblowsky takes the argument a step further by claiming that in the Lurianic Kabbalah man essentially becomes the savior of God:

> Israel's exile became meaningful because it was seen as a participation in the profounder exile of God, and God Himself required Israel's active participation in the redemption of Himself and His people. It is not surprising that in this kabbalistic system the personality of the messiah played a relatively minor role. He was not so much a redeemer as a sign and symbol that the redemptive process has been achieved. In fact, the messianic doctrine of Lurianic Kabbalah comes close to the structure of an evolutionist scheme.[29]

The Lurianic Kabbalist could not retreat into his own private world and passively await the Messiah. He had to participate in a cosmic millennial drama in which his every action counted. The Lurianic Kabbalah was the first Jewish theology that envisioned perfection as a future state, not as a forfeited ideal past, and as such it made a direct contribution to the idea of progress emerging in the West.

This same sense of urgency motivated many Christian millenarians. In Acts 3:19–21 Peter speaks of "the times of restitution of all things." Many people including Bacon took this to mean that humans would be restored to the state of Adam in the Garden of Eden, perhaps not completely but to a large degree. In his *Advancement of Learning,* Bacon promised that learning would lead to "a restitution and reinvesting (in great part) of man to sovereignty and power . . . which he had in the first state of creation."[30] Sir Thomas Browne struck a similar optimistic note in his *Pseudodoxia Epidemica* when he counseled that although "now our understandings . . . [are] eclipsed, as well as our tempers infirmed, we must betake ourselves to reparation, and depend upon the illumination of our endeavours. For thus we may repair our primary ruins."[31] Even John Donne, who viewed the modern fragmented world with such pessimism, states that "Our busineese is to rectifie/Nature, to what she was."[32]

The early Christian theologian and philosopher Origen (c. 185–254), whose belief in universal salvation reemerged in the seventeenth century, had dismissed a literal view of scripture as "silly." As he wrote in *On First Principles*: "And who is so silly as to believe that God, after the manner of a farmer, 'planted a paradise eastward of Eden?' These are figurative expressions which indicate certain mysteries through a semblance of history and not through actual events."[33] The answer to Origen's rhetorical question fifteen centuries later was that many people were just this silly. Peter Harrison makes a compelling case for recognizing biblical literalism from the Reformation forward as an indispensable factor in the Scientific Revolution. The notion that the Bible was the font of all knowledge was a common opinion from early Christianity onward, but the peculiarly literal way this was understood in the early modern period was new and unprecedented. The premise of Lambert Daneau's *The Wonderful Workmanship of the World* (1578) was that "general naturall Philosophie . . . is chiefly to bee learned out of holy Scripture."[34] Levinus Lemaius, author of *An Herbal for the Bible* (1587), assures the reader that scripture is "most exquisitely . . . furnished with the entire knowledge of all things naturall: and not ignorant in anie kind of learning or discipline."[35] Christian Knorr von Rosenroth, emphasized this point when he describes the Old Testament as a "gold mine" and "treasure chest" of knowledge:

> And if we examine the golden writings of the Old Testament, what do we find there but the deepest goldmine of all the arts and sciences and a treasure chest, in which the most precious gems of philosophy and the immense riches of the law and, first and foremost, all the treasure of divine and salvific wisdom? . . . .they display the purest mirror of all the virtues and vices, and . . . contain the fountain from which we can drink the most precise principles of politics and economics. If only I could demonstrate more fully how those remarkable books contain in their simple written letters not only the secrets of nature but also the secrets of civilization, to say nothing of the secret prophecies![36]

In his *Conjectura Cabbalistica* (1653) Henry More provided an ingenious analysis of the first three chapters of Genesis to show that Moses anticipated Descartes.[37]

This view of the Bible as the repository of all knowledge was supported by the Humanists and Reformers of the sixteenth and seventeenth

Frontispiece. John Parkinson, *Paradisi in sole paradisus terrestris*. . . . London, 1629. (Wellcome Library, London)

centuries, who encouraged reading the Bible as literal history. Books and works of art routinely included illustrations of biblical stories and diagrams of the Ark, the Tower of Babel, and the Temple, as well as maps pinpointing sites mentioned in the biblical text, reinforcing the conviction that the Bible offered an accurate historical account of the world. The English Geneva Bible (1615), for example, shows the location of the Garden of Eden, and Raleigh's *Historie of the World in Five Bookes* (1614) contains an illustration of the Ark after it landed on Mt. Ararat. The title page of John Parkinson's *Paradisi in sole paradisus terrestris* (1656) is one of many depictions of the Garden of Eden published in the early modern period, all of which inculcated the idea that Eden was a real place with a real history.[38] Jim Bennett and Scott Madelbrote have shown that the biblical stories of the Garden of Eden, Noah's Ark, the

Tower of Babel, and Temple of Solomon were mentioned so frequently in the early modern period because they were thought to provide readers with exemplary moral tales about real people and real events that, while warning of the dangers of human pride, offered hope that the fallen world might be restored to its original condition. The story of the Fall, for example, revealed how pride and sinfulness led to the corruption of man and nature. Instead of simply plucking fruits from the trees to gain his daily bread, man was forced to labor by the sweat of his brow and wrest food from a grudging earth. This was a tragic tale to be sure, but one that led many early modern readers to conclude that improvements in morality might lead to improvements in agriculture, and that both might lead man back to paradise, where Adam had dominion over nature. Joseph Glanvill viewed the research undertaken by members of the Royal Society in exactly this light. He claims it had contributed to "the accelerating and *bettering* of *Fruits*, *emptying mines*, *drayning Fens* and *Marshes*." Because of their endeavors "*Lands* may be *advanced* to scarce credible degrees of improvement, and innumerable other *advantages* may be obtained by an *industry* directed by *Philosophy* and *Mechanicks*." All these activities were instrumental in "*captivating Nature*, and making her *subserve* our *purposes* and *designments*" with the laudable goal of restoring "the *Empire* of *Men* over *Nature*."[39] In the minds of Glanvill and his contemporaries, God was "the originall, and first Husbandman." If the image of God was to be restored to man, there was no better way to do it than by imitating what God did by restoring the earth to its pristine condition. The image of Christ as a gardener was popular in the early modern period and based on the account of Jesus's first appearance to Mary Magdalene, when she mistook him for a gardener (John 20:15). Such depictions are a good example of how the preoccupations of biblical readers influence the way the biblical message is interpreted, for this image of Jesus ignores the fact that the offerings of Cain, the farmer, were rejected by God in favor of the gifts of Abel, the shepherd.

Millenarians did not just talk about restoring the Garden of Eden; they set about doing it. Some of the more radical ones went so far as to wear the original clothes of Adam and Eve, or the lack thereof, and they returned to the vegetarian diet of the Garden. To illustrate how difficult this was for one determined but carnivorous Englishman, we have only to look carefully at the frontispiece of *Paradisi in sole paradisus terrestris*, mentioned earlier, to discover a new and marvelous species, the

Close-up of the "Vegetable Lamb." Frontispiece. John Parkinson, *Paradisi in sole paradisus terrestris*. . . . London, 1629. (Wellcome Library, London)

"vegetable lamb." This exotic animal conveniently grew and propagated like a plant, grazing on the grass around its stem. Practically oriented millenarians developed new techniques of husbandry and designed new agricultural tools. Walter Blith's *English Improver Improved*, first published in 1649 and in an emended and enlarged edition again in the same year as well as in 1652 and 1653, was one of the most important handbooks on agricultural improvements. Blith illustrates a variety of tools for digging and ploughing in an effort to educate farmers and stop the English practice of plowing too deeply, which he claimed reduced the fertility of the soil and wasted seed. The millenarianism of the author is revealed in the frontispiece, which depicts Royalist and Parliamentarian troops lined up for battle. But some of the soldiers have defected and are beating their weapons into ploughshares. Blith's goal was the "Reducement of Land to pristine Fertility," and he criticized the sloth and debauchery of his countrymen for delaying it: "were ingenuitie the Fashion of the Times, This Kingdom would be the Paradise of the World."[40] Samuel Hartlib (1600–1662), the central figure in the group of "Pansophic" reformers previously described, arranged for the publication of *The Reformed Husband-Man; Or a Brief Treatise of the Errors, Defects, and Inconveniences of our English Husbandry*. This was one aspect of Hartlib's many-faceted and untiring efforts to improve the world and bring on the millennium. Tree planting was another favorite activity of millenarians. The connection between agricultural reform and spiritual progress is highlighted by two books by Ralph Austen also published by Hartlib in 1653, the first, *A Treatise of Fruit-Trees*, and the second, *The Spiritual Use of an Orchard*. While Warden of Wadham College John Wilkins (1614–72), a founding member of the Royal Society who became Bishop of Chester, enlarged and improved the gardens, installing transparent beehives for the delight of his guests and watering devices with multiple spouts based on designs from the *Pneumatics* of Hero of Alexandria. Like many of his contemporaries, Wilkins experimented with exotic crops from the Americas and the Far East.[41] He had a patch of Indian wheat in his Oxford garden, and Robert Boyle wrote to Hartlib about the "peculiar Flowers" cultivated there.[42] Given the interest in recreating the Garden of Eden, it is no coincidence that six of the most famous botanical gardens in Europe were established during the sixteenth and seventeenth centuries at Padua, Leiden, Montpellier, Oxford, Paris, and Uppsala.[43] The rationale behind millennial thinking about agriculture was that the earth had to be cultivated to reverse both the curse occasioned by the Fall and the subsequent injury to the land caused by the Flood. Uncultivated land was a sign that God's will had not been fulfilled because vacant land was looked upon as "a deformed chaos."[44] The failure of native peoples to cultivate their territory became one of the primary justifications for colonization and a rationale used by Locke in his *Second Treatise on Government* for why European settlers had a right to expropriate virgin territory, notwithstanding the Native Americans who considered it sacred.[45]

The hopeful message that millenarians took from their conviction that the Garden of Eden might be restored was reinforced by their interpretation of the Flood. The fact that Noah was a righteous man, who had retained enough of Adam's primordial wisdom to construct the Ark, assemble the animals, and bring this vast menagerie of men and beasts to a new world and a new beginning, gave millenarians hope that humans were less damaged by the Fall than many people imagined. There was a long tradition in Christianity of Noah as a type of Christ and the Ark as

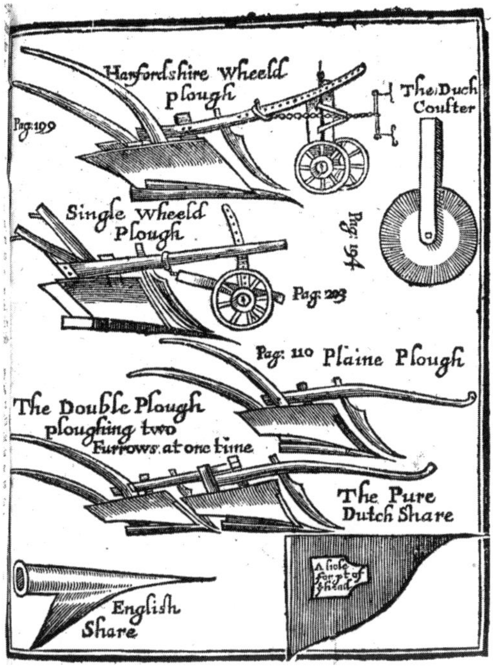

Types of ploughs. From Walter Blith's *English Improver Inproved.* London, 1652. (Wellcome Library, London)

an allegory of salvation. This positive view of Noah was current in the early modern period and offered a model of how fallen men might cultivate their innate intelligence, knowledge, and skills and regain some, if not all, of Adam's prelapsarian wisdom. Because of his skill in building and navigating the Ark, Noah was envisioned as an astronomer, shipwright, navigator, mathematician, and architect as well as a naturalist, familiar with the habits and diets of every animal species.[46] Taking the story literally raised a host of practical questions: Where did the flood waters come from? Where did they go? How was the earth changed by the Flood? How could the Ark house all the animals? How did Noah collect them, get them onto the Ark, and feed them once there? Did they procreate, and if they did, how did Noah deal with the offspring? Calvin had been disgusted by the thought of all the manure that must have been produced and wondered how Noah kept the Ark clean.[47] Such questions inspired John Wilkins to make detailed calculations about these matters. He determined that the carnivores on board would require 1,825 sheep to

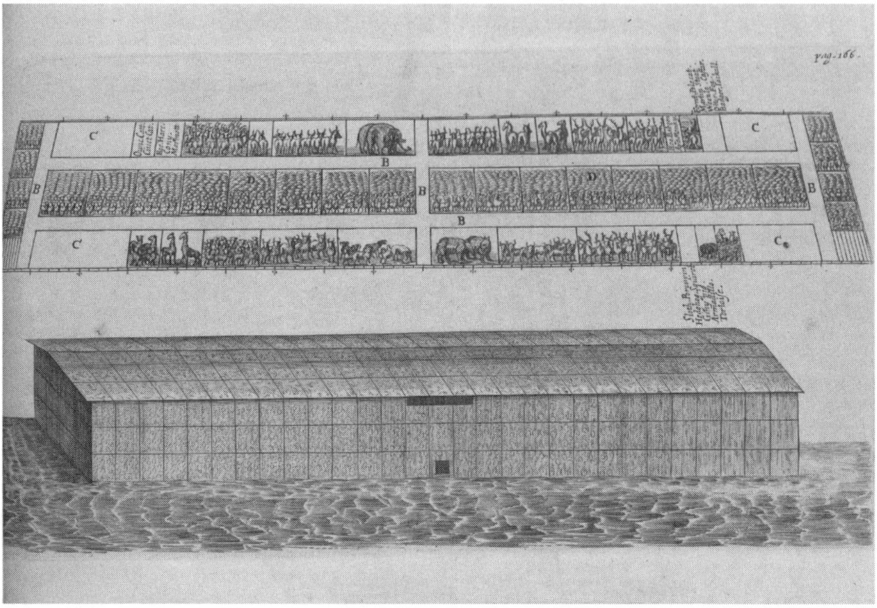

Noah's Ark. From John Wilkins, *An Essay towards a Real Character, and a Philosophical Language*. London, 1668. (Wellcome Library, London)

eat during their year on the water, while enough hay for the ruminants could be collected and stored on the middle deck. Wilkins's design of the Ark is as meticulous in its detail as his calculations.

Noah's ark became a prototype for the encyclopedic cabinets of curiosities and collections of plants, animals, and natural objects that proliferated throughout Europe in the early modern period. The story inspired people to collect and classify specimens in an effort to regain the kind of natural knowledge lost through the Fall and with it the dominion over the natural world possessed by Adam. Paula Findlen considers these collections and the museums in which they were housed as forerunners of London's Royal Society and the French Academy of Sciences. Thus, biblical precedents when taken literally were influential in shaping the early modern scientific imagination.[48] Even the cautionary tale of the destruction of the Tower of Babel was given a positive spin in early modern texts because it suggested to many naturalists and philosophers that the Tower might be restored and people reunited through the creation of a "universal" or "natural" language in which words would be such clear reflections of things that confusion and misunderstanding would be impossible.[49]

The British were particularly interested in language reform during the seventeenth century and it has been argued (and denied) that this interest was stimulated by the work of Jan Amos Comenius.[50] Expelled from his native Bohemia during the Thirty Years' War, Comenius traveled through Europe inspiring those he met with his zeal for educational and social reform. His two staunchest supporters, Samuel Hartlib and John Dury, joined him in promoting "Pansophy" and the reformation of all the arts and sciences. Language reform played a central role in their schemes, for like many others Comenius, Hartlib and Dury were convinced that the misuse and misunderstanding of words were the principle causes of dispute.[51]

Comenius's arrival in England in 1641 may have inspired the English to think about a natural language, but there were factors that contributed to the great interest in language reform as well. One of the most important of these was war, both the Thirty Years' War and the English Civil War. Confronted by the grim consequences of intolerance many people concluded that religious conflicts had more to do with the misuse of language than with fundamental differences of belief. To overcome this, they tried to devise languages in which words would be so clearly related to things that no one could dispute their meaning. In his *Universal Character* (1657), for example, Cave Beck hoped the artificial language he devised would help to propagate "true religion in the world." William Petty, later an energetic and versatile member of the Royal Society, turned his attention to a dictionary of difficult words because he believed that the conflict between Catholics and Protestants rested largely on the misunderstanding of words like "God," "Devil," "Heaven," "Hell," "Catholic," and "Pope."[52] John Wilkins hoped his natural language would be useful in settling religious disputes.[53] Leibniz tried to devise a language that could be manipulated mathematically so men could "calculate" their differences away.[54]

The belief that a simple, clear, and natural way of speaking would overcome all differences and provide the foundation for the advancement of learning was a guiding principle of the Royal Society. In his history of that institution, Thomas Sprat wrote,

> [it] has been, a constant Resolution, to reject all amplifications, digression, and swellings of style: to return back to the primitive

> purity and shortness, when men deliver'd so many things, almost in an equal number of words. They have extracted from all their members, a close, naked, natural way of speaking; positive expressions; clear senses; a native easiness: bringing all things as near the Mathematical language of Artizans, Countrymen, and Merchants, before that, of Wits, and Scholars.[55]

The vision ascribed by Sprat to the members of the Royal Society was of a universal philosophy that surmounted differences of birth, nationality, and religion; the aim was "not to lay the foundation of an English, Scotch, Irish, Popish or Protestant philosophy, but a philosophy of mankind."[56] However, as Christoph Lüthy notes, this was a difficult undertaking. Seventeenth-century science was as "confessionalized" as religion, although the growing emphasis on experiment and fact-finding helped to build bridges across denominational divisions by separating observational facts from theological frameworks.[57] Nevertheless there were clearly touchy issues in philosophy and theology that could and did flare up in the Royal Society. For example, the controversy between Seth Ward, Savilian professor of geometry at Oxford, and John Webster, a follower of Paracelsus and Boehme, and the debate between Henry More and Robert Boyle over hydrostatics, while seemingly scientific, had theological implications that threatened to undermine the civility of the participants.[58] Controversies like these, together with more acrimonious ones dealing directly with religion, made the idea of a natural language seem so desirable and beneficial that it became a common feature in contemporary utopias. In his "history" of the Sevarites, for example, Denise de Vairasse ascribes the "innocence," "politeness," and "happiness" of these imaginary natives of Australia to the clarity of their language:

> by the care which is taken to teach every Person the Principles of Grammar, they speak better, and express themselves more clearly, than any Nation in the world; from whence we may conclude, that they surpass us as much in the beauty of their language, as in the innocence and politeness of their manners, and that they are not, excepting only the article of Religion, the Happiest People on Earth.[59]

The dream of discovering or creating a perfect or natural language turned out to be just that, an unrealizable dream. But although they failed, the work of the language projectors had great significance for subsequent history. On the positive side, the idea that all men shared certain common notions that could be expressed in a common language fostered the ideals of democracy and equality and consequently provided the basis for political freedom, individual rights, and toleration. In this sense the language projectors helped to set the stage for the progressive political and social ideas of the Enlightenment. An extraordinary degree of optimism was fostered by the ideal of a natural language because it seemed to offer a mechanical and therefore infallible method for solving disputes as well as for discovering new truths. "Method" becomes a key word in the seventeenth century and, indeed, in later centuries. The faith in method so characteristic of language projectors and natural philosophers encouraged the optimism and belief in progress that went hand-in-hand with the development of modern science.

The schemes of the language projectors are important for another reason as well. They were theologians, naturalists, and philosophers deeply interested and involved in the philosophical and scientific debates of their day. Many were members of England's Royal Society and other national and regional scientific institutions. Yet their work exhibits traces, and in some cases large deposits, of what can best be described as esoteric ways of thinking. A number were millenarians, others were vitalists and subscribed to esoteric forms of thought such as Hermeticism and the Kabbalah. Even those who expressly repudiated esoteric thought were not entirely free of esoteric influences. Taking their ideas seriously therefore qualifies the characterization of the seventeenth and eighteenth century as the period in which science flourished because a mechanical, atomistic, or corpuscular worldview triumphed over outmoded esoteric, magical, and occult thinking. Their work provides evidence that calls for a more nuanced view of the Scientific Revolution, one that includes the role of esotericism in its unfolding. This is a subject to which we will return in chapter eight.[60]

The story of Solomon's Temple was another key text that resonated positively in the minds of early modern biblical literalists, for the Temple was both a repository of wisdom and an architectural masterpiece embodying the principles of divine geometry.[61] The fact that Solomon, like Noah, was renowned for his wisdom long after Adam and Eve were expelled from the Garden of Eden provided another positive example of how the loss of Adamic knowledge might be overcome through diligent study and hard work.[62] Solomon was in fact the very model of a natural philosopher:

> And God gave Solomon wisdom and understanding exceeding much. And Solomon's wisdom excelled the wisdom of all the children of the east country, and all the wisdom of Egypt. And he spake of trees, from the cedar tree that is in Lebanon even unto the hyssop that springeth out of the walls; he spake also of beasts, and of fowl, and of creeping things and of fishes (1 Kings 4:29–33).

Solomon's temple became an archetype for scientific institutions like the Royal Society.

The biblical stories of the Garden of Eden, the Flood, The Tower of Babel, and Solomon's Temple fueled the millenarian dreams and utopian speculations of natural philosophers in the early modern period. Unlike today, when a literal interpretation of the Bible characterizes those who oppose science, in the early modern period a literal reading produced the opposite effect, encouraging scientific inquiry by providing the model of a perfect past that might be restored through human effort and ingenuity and lead to an even more perfect future. Thus literalism, which was thought by Galileo to impede scientific research, could do the opposite. Peter Harrison is eloquent in making the case that Protestant literalism was an essential factor in the rise of modern science. It was biblical literalism and the rejection of allegorical interpretation that freed the study of the natural world from religion:

> This insistence on the primacy of the literal sense had the unforeseen consequence of cutting short a potentially endless chain of reference, in which word refers to object, and object refers to other objects. The literalist mentality of the reformers thus gave a determinate meaning to the text of scripture, and at the same time precluded the possibility of assigning meanings to natural objects. . . . In this way the study of the natural world was liberated from the specifically religious concern of biblical interpretation, and the sphere of nature was opened up to new ordering principles.[63]

As the foregoing discussion has tried to show, taking religion seriously has led to the discrediting of the grand narrative prevailing up to the 1960s that described the Scientific Revolution as going directly from Copernicus to Newton via Kepler and Galileo. During the past forty years astronomy and physics have been demoted from their primary position as the fundamental sciences of the Scientific Revolution, and the emphasis on mathematics as the single most essential aspect of modernity has been rejected, thus allowing the inclusion of developments in other areas of natural philosophy.[64] For example, Aristotelianism was brought back into the picture by Charles Schmitt, who, along with Mordecai Feingold, reevaluated the role of universities and the study of Aristotle in the Scientific Revolution.[65] In addition, subjects previously marginalized such as alchemy, astrology, magic, Neoplatonism, natural history, and antiquarianism were added by a long list of scholars beginning in the anglophone world with Frances Yates. Medicine and biology have also been reintroduced into the picture and recognized for what they contributed to the Scientific Revolution. As a consequence, a new picture of the Scientific Revolution has emerged. Historians have increasingly stressed the fact that not only was early modern "natural philosophy" much broader than modern science because it included theological issues involving creation, providence, the immortality of the soul and its relation to the body, and the effect of the Fall on human rationality, but in many instances these theological issues fostered scientific development. Consequently it is impossible to distinguish between religion, magic, and science in the early modern period or to make a clear distinction between occult and scientific mentalities, as Brian Vickers tried to do.[66] It is also difficult to argue that the so-called "mechanical philosophy" became the predominant paradigm.[67] Vitalism did not disappear, and "spiritual" forces appear in even the most seemingly mechanical of worldviews.[68] Betty Jo Dobbs, the historian whose work demonstrated the centrality of alchemy in Newton's thought and its influence on his physics, noted that this new history of science has plunged historians into strange encounters with figures we thought we knew but now have to struggle to comprehend:

> we look at them a little more closely and discover to our astonishment that our intellectual ancestors are not like us at all: they do not see the full implications of their own work; they refuse to believe things that are now so obviously true; they have metaphysical and religious commitments that they should have known were unnecessary for a study of nature; horror of horrors, they take seriously such misbegotten ideas as astrology, alchemy, magic, the music of the spheres, divine providence, salvation history. We become most uncomfortable and begin to talk about Copernicus as "conservative" or "timid," terms that hardly fit the commonsense concept of

> revolutionary. Or we talk about Kepler as a "tortured mystic" or a "sleepwalker" or a "split personality."[69]

D'Alembert experienced this kind of difficulty in understanding Bacon: "After having burst so many irons, this great man was still held by certain chains which he could not, or dared not break."[70] D'Alembert is not alone; we have similar problems: How could John Dee have ever imagined he could communicate with angels? Why did ostensibly sane and educated men believe old women had sex with the devil? What made Robert Boyle waste money subsidizing an account of a poltergeist? Not only is the mental landscape of early modern natural philosophers foreign territory in many respects, but the greatly expanded view of what counted as natural philosophy has taken historians of science into new territory as well, where religion, magic, and science mixed together in unexpected and fruitful ways.

CHAPTER 8

ESOTERICISM AND THE SCIENTIFIC REVOLUTION

The term "Christian Europe" or "The Christian West" is incapable of conveying the multiple religious, occult, and magical beliefs of Europeans and Americans in the past. This realization has led to a fundamental reevaluation of the nature of Western religious practice and belief by showing that religious identities were never fixed along simple denominational lines but constructed out of a plurality of available religious, scientific, and philosophical discourses often originating in esoteric sources. As a field of study esotericism has therefore come into its own in recent years. Encompassing magic, astrology, alchemy, Hermeticism, the Kabbalah, and a host of subjects that were previously relegated to the rubbish bin of history, the study of Western Esotericism has challenged and continues to challenge mainstream historiography. Scholars from across the disciplinary spectrum have become increasingly aware of the way esoteric currents of thought have shaped Western history, science, religion, philosophy, anthropology, sociology, art, and architecture, as well as the academic study of these disciplines. Consequently, the study of esotericism has proven its worth not only by providing a more nuanced picture of major developments and events from medieval to modern times but also by revealing the complex strands of esotericism in the thought of key historical figures.[1]

In one of the latest studies of esotericism, Kocku von Stuckrad argues that esotericism must be seen as a structural element of Western culture that played and continues to play a key role in shaping European identities and constructing fields of knowledge. In the course of substantiating these claims, he provides additional evidence to undermine Weber's thesis, calling into question both the notion of Western identity as rational, enlightened, and scientific and the idea that the last three-hundred years have witnessed an irreversible process of secularization. Religion has not disappeared in the West because religion as a singular entity never existed. Thus the notion of a "Christian" Europe is a misnomer. Plurality of religion has always been the norm, for it is only by defining the deviant other that any group defines itself. Three assumptions underlie this analysis: first, religious pluralism and the existence of alternatives represent the normal situation in Europe, not an exception; second, Western culture has always been characterized by critical reflection on religious truth claims and different systems of knowledge and social organization; and third, competing ways of attaining knowledge of the world explain and confirm the crucial role of esotericism in Western discourse. In this formulation, esotericism is not, as many nineteenth- and twentieth-century scholars (Scholem, Corbin, Eliade, Jung) imagined, some kind of subculture that provided an alternative to the religious mainstream, whether that be Judaism, Christianity, or Islam. It represented a different approach to knowledge and existed among the adherents of all religious traditions, one that privileged the experiential dimension of religious life and was predicated on the idea that perfect knowledge of God and the cosmos is possible.

Von Stuckrad juxtaposes two basic traditions in Western thought, one characteristic of esoteric knowledge and based on the belief that direct experience of divine and transcendental truths is attainable versus the more skeptical view that it is impossible to attain knowledge beyond rational demonstration. Cornelius Agrippa provides an example of the first when he insists that to have true understanding and knowledge

> it is needful to have a higher spirit to judge and discern, which is not given us by men, nor by flesh and blood, but is given from above by the father of light, for none without this light can truly speak any godly thing. And this light is God's word, by which all things are made, giving the light to every man that cometh into this world, and giving them power to be made the sons of God.[2]

Agrippa's emphasis on revelation as the true source of knowledge is radically different from the kind of skeptical approach taken by Montaigne or Charron. To a large extent, these two positions were encapsulated in the nominalist/realist debate of the Middle Ages. Esotericists were realists deeply indebted to Platonic and Neoplatonic traditions (and those currents allied with them such as Gnosticism, Hermeticism, Kabbalah, and Pansophism). Language was a key issue in this debate: those in the realist camp believed that language reflected the cosmos. Words in this view were identical to things, and by decoding words it was possible to understand the essential nature of created entities. This way of thinking provided the rationale for the emblematic view of the world discussed in chapter one. Opposed to this view were the nominalists, who denied that names or words had any ontological status; they were simply arbitrary human constructs. Thus while language was an invaluable tool enabling human beings to gather and sort information, this information would always be partial, imperfect, and limited to the natural realm of experience and the senses. Where nominalists insisted on the strict separation between God and man and between the divine and the natural worlds, realists blurred the line between religion, magic, and science, offering a kind of "linguistic gnosis" that put humans on a par with God. Alchemical writings provide many examples of this approach to knowledge. Von Stuckrad provocatively suggests that our modern obsession with codes and decoding, especially in regard to genes, is a modern outgrowth of this ancient and esoteric view of the universe as "readable" and of human beings as the ultimate readers of texts. The idea that humans can possess divine knowledge to the point that they become divine appears so radically at odds with notions of human nature traditionally associated with monotheistic religions as to be almost unbelievable. Yet esoteric currents of thought within monotheistic traditions continually broke down the barriers separating the human from the divine. Once again the category of esotericism helps to illuminate important aspects of and conflicts within Western thought that would otherwise seem inexplicable.[3]

Esoteric thought was thus a normal element of European culture. It appears in somewhat different guises in religious and intellectual circles across Europe and through the ages, but all these groups were united by certain "shared passions" for attaining absolute knowledge in various forms (magic, alchemy, astrology, sacred history, universal languages). These shared passions for esoteric knowledge led to a kind of "interconfessionalization," whereby certain Jews, Christians, and Moslems actually had more in common with each other than with their own religious group, a phenomenon that should be recognized as an important aspect of European history.[4]

In the same way that esoteric thought cannot be separated from religion, so too is it impossible to separate esotericism from science and cultural innovations in general. Astrology, for example, was so closely linked to mathematics, medicine, philosophy, and natural philosophy in the medieval and early modern periods that it distorts the historical record to view it as a pseudo-science. For this reason, it is impossible to disentangle "real" from "pseudo-science" in the work of the Elizabethan mathematician and magus John Dee (1527–1608/9). Dee's interest in mathematics cannot be pried loose from his esoteric pursuit of absolute knowledge, a pursuit that led him into alchemy and astrology and encouraged him to initiate conversations with angels. His angelology was part of his interest in natural philosophy, not antithetical to it.[5] The same can be said of Newton. The attempt to separate his "good" physics" from his "bad" theology and alchemy is not possible.[6]

The emergence of the field of estoericism was accelerated by the publication of Frances Yates's *Giordano Bruno and the Hermetic Tradition* (1964), in which she argued for the importance and widespread influence of the "Hermetic-Cabalist" tradition on the arts, philosophy, and science of the Renaissance and early modern periods. By Hermeticism Yates meant the treatises in the *Corpus Hermeticum*. Although these texts were actually written between the second and fourth centuries CE, they were thought to be extremely ancient because they were ascribed to Hermes Trismegistus, or Thrice-Great Hermes, reputed to be a contemporary of Moses and the founder of alchemy in the West. So revered was the figure of Hermes Trismegistus that he is portrayed on the beautiful black and white inlaid marble floors of the cathedral in Siena. This very fact is an indication of how flexible the notion of "Christianity" was at the time. Given their provenance, the Hermetic texts were considered a source of the *prisca theologia* ("first theology") that had been imparted by God to Adam in the Garden of Eden and again, according to some commentators, to Moses on Sinai. Europeans in the early modern period, especially ones with an ecumenical vision, were anxious to rediscover this "first philosophy" because they believed it would provide a philosophy and theology that everyone could accept—Christians of all denominations, along with pagans, Jews, and Moslems.[7] Cosimo di Medici purchased a Greek manuscript of the *Corpus Hermeticum*. He considered it so important that he ordered the chief philosopher at his Florentine Platonic Academy, Marsilio Ficino, to put aside his translation of Plato's dialogues and work on a translation of the Hermetic texts instead.

Yates shared Cosimo's conviction that the Hermetic texts were of crucial importance, although for very different reasons. She was convinced that modern science originated in the new and optimistic view of human nature that emerged as the Florentine Neoplatonists in Cosimo's Academy read and absorbed Hermetic texts. She believed this optimism was carried into the seventeenth century through the subterranean channels of esoteric philosophy, contributing to the anthropological revolution described in chapter five. The idea that man could change his environment for the better and harness the powers of nature to his own advantage had its roots in the magical world of Renaissance Hermeticists, and the twin concepts of progress and reform that became the hallmarks of modern science emerged from the grandiose schemes of Renaissance magi, not from the patient accumulation of scientific evidence and scientific theories. In Yates's view the *Rosicrucian Manifestos* of the early seventeenth century were perfect expressions of this exhilarating view of human potential and prowess that made the Scientific Revolution possible. With their call for the "Universal and General Reformation of the whole world" and their conviction that creation can be brought back to the state in which Adam found it, *The Rosicrucian Manifestos* provided a bridge between Renaissance Hermeticism and modern science. On the basis of this evaluation of their importance, Yates suggested the word "Rosicrucian" should enter the vocabulary of serious historians to describe the kind of activist, reforming mentality paving the way for modern science.[8]

Yates's reputation was at its height during the 1960s and early 1970s. But it wasn't long before a "backlash" set in. Charles Schmitt believed her stress on Renaissance occultism and Hermeticism was exaggerated and reinforced a major misapprehension in the history of science, namely that "modern science began when Aristotle's authority had been replaced." Schmitt was convinced that the concentration on the mathematical and quantitative aspects of sixteenth- and seventeenth-century science had led to a distortion of what actually occurred during the Scientific Revolution, making it impossible to understand the persistence of Aristotelianism in the universities, the evolution of other sciences, particularly biology, and even the place of esotericism in the rise of science.[9] Schmitt's multifaceted analysis of the Scientific Revolution has had more impact than Yates's Hermetic hypothesis. Yet the fact that Schmitt recognized a place for esotericism in the rise of modern science is largely attributable to Yates and reveals the effect she has had in making it a legitimate subject of study.

Yates's stress on Hermeticism and occultism and her notion of a "Rosicrucian Enlightenment" received further serious criticism from Brian Vickers. The idea that good science could come out of bad occultism clearly affronted Vickers.[10] He was particularly resistant to the idea that there could be any connection between a "scientific" and an "occult" or "esoteric" mentality. In his view esoteric and scientific methodologies are so radically different that any influence of one upon the other is impossible. In esoteric systems categories are determined a priori; they are not neutral as they are in scientific systems but analogical, symbolic, evaluative, and hierarchical. Where the scientist aims at causal action, the esotericist only acts symbolically.[11] The problem for Vickers then becomes "to understand how . . . men were able to operate simultaneously with two traditions that have become generally recognized as incompatible since, say, the first generation after Newton."[12] Vickers suggests it may be a mistake to try to see consistency in Newton; instead of one Newton, there may indeed be two, a successful scientist and a failed esoteric alchemist.

Vickers's analysis fails to explain why so many thinkers in the early modern period clearly belonged in both the esoteric and the scientific camp, a point made repeatedly by many historians in recent years.[13] If it were indeed so easy to distinguish between science and esotericism, how does one explain Bacon's *New Atlantis* or other aspects of his hardly straightforward philosophy, and how does one account for the mixture of scientists and estoericists (alchemists, astrologers, Paracelsians, Hermeticists, and Kabbalists) who made up the membership of the Royal Society?[14] Walter Pagel devoted a lifetime to illustrating the intricate connections between science, magic, and religion in the work of such figures as Paracelsus, van Helmont, and William Harvey.[15] Richard Popkin also demonstrated the impossibility of isolating science from theology in the early modern period.[16] Richard Westfall and Betty Jo Dobbs have shown that Newton's alchemical studies influenced his physics—and one should note that Westfall was a convert to this conclusion fairly late in his academic career, as one can see from his earlier discussions of Newton.[17] Nor were Boyle, Locke, and Leibniz immune to the charms of esotericism. As we have seen, Boyle may have been instrumental in collecting cases of "second sight" in the Scottish Highlands.[18] Earlier in his life he had shown great interest in collecting empirically verifiable accounts of witches and spirits. Boyle also had lifelong interest in alchemical transmutation. For a considerable period he was duped into sending money and gifts to an international, secret sect of alchemical adepts. Its members included one "Sabitieli," who possessed a powder that could coagulate water into a transparent stone; a "Polish philosopher," who caused plants to flower and fruit in the space of two hours; and (my favorite) a "Chinese gentleman" named "Pursafeda," who exhibited flasks containing a developing homunculus, a five-month-old foal, and a fox. These examples should make us aware of how fluid the line between magic and science was at the time Boyle lived and in Boyle's own mind.[19] Boyle, Locke, and Newton were all deeply involved in alchemy and alchemical experimentation, an indisputable fact leading several scholars to suggest that Newton's mental breakdown in the 1690s and Boyle's chronic ill-health may have been caused or aggravated by their contact with mercury, a key ingredient in alchemical experiments.[20] Locke's medical interests during his Oxford period led him to read widely in the occult medical tradition represented by Fernel, Cardano, Paracelsus, Campanella, Sennert, and J. B. van Helmont. In the 1660s Locke read the alchemical works of Basil Valentine, and in the early 1690s his duties as Boyle's literary executor "relit the Hermetic flame," which had dimmed under the medical influence of Sydenham. As one scholar comments, "Locke's journals remind us that the natural historical impulse in early modern thought owes as much to magic as to science, as much to della Porta as to Bacon."[21] For all their apparent differences Locke and Leibniz shared an interest in alchemy and esotericism. As I have argued, Leibniz's mature philosophy was influenced by the Kabbalah, and analogical thinking was a tremendously important aspect of his work. That Locke and Leibniz were both good and loyal friends of Francis Mercury van Helmont (1614–98), an alchemist and Kabbalist, and that both owned and read his Kabbalistic works shows how anachronistic our modern categories are when applied to this earlier period.[22]

The great polymath and Jesuit Athanasius Kircher (1602–80) provides another example of someone whose interests obscure any boundary separating religion, magic, and science. At the age of twenty-eight he began teaching mathematics, Hebrew, and Syriac at Würzburg. Later in life he became fascinated by Egyptian hieroglyphics, a passion that lasted throughout his career. He loved mechanical inventions and designed a magnetic clock, various kinds of automatons, and a megaphone. He was also interested in fossils and volcanos. When Mt. Vesuvius erupted in 1638 he had himself lowered into the cone for better observation. He was interested in telescopes, sunspots, and magnetism, the subject of his first book *Ars Magnesia* (1631). He originated and oversaw one of the greatest collections of curiosities and marvels in the early modern period, the Kircher Museum in the College of Rome. At the same time he accepted astrology, claimed to have performed palingenesis (resurrecting plants and animals from their ashes), and believed in mermaids, gyphons, the geocentric world, and the Hermetic tradition, rejecting Copernicus.[23] Although considered a scientific genius and true man of the Renaissance in his day, Kircher has been largely ignored until recently when scholars began to appreciate how important his work is if we are to develop a real understanding of the mental, spiritual, and scientific lives of natural philosophers in the early modern West.[24] On the basis of Kircher alone, not to mention all the other figures like him, a distinction between an occult and scientific mentality simply does not fit the facts in the sixteenth, seventeenth, or well into the eighteenth century. Paolo Rossi makes this point when discussing Jean Baptiste Robinet's four volume work *De la nature* (1761–68) published almost a century after Kircher's death:

> As if reinvigorated . . . some of the themes most characteristic of the "hermetic tradition" and the "magical" vision of the cosmos spring back to life: vitalism, universal sympathy, the analogy between beings, the ladder of nature, the presence of the All in even most minute and apparently insignificant aspects of reality . . . many of his pages anticipate themes that would have singular success in the romantic philosophies of nature from Goethe to Schelling and beyond.[25]

To carry the issue into the modern period, how can the radical distinction between scientific and occult methodologies explain the continuation of a priori analogical and symbolic thinking among scientists well into the twentieth century? On what other basis than analogical thinking did Social Darwinists and biological determinists like Broca, for example, justify their racism, sexism, and class bias? Turning the question around, scholars have shown that what were once considered occult and hence non-scientific aspects of Bruno's thought; for example, his interest in images and visualization can now be seen as remarkable insights into some of the issues facing modern science.[26] The same thing can be said of Leibniz, whose monads are more in tune with modern concepts of matter as fields of force than the supposedly progressive atomism and mechanical philosophy of the Scientific Revolution.

Thus, Yates's hunches about the importance of esotericism in fostering science proved to have far more substance than her detractors imagined. Mounting evidence confirms that major thinkers in the scientific pantheon were influenced by esoteric theories. Rosicrucianism, for example, was more widespread than Yates realized, appearing in Scotland, Germany, and the Baltic among a great number of intellectuals and natural philosophers.[27] Consequently, it has become less and less tenable to separate good science from bad occultism. This was the basic conclusion reached by the panelists who assembled at the Folger Library in 1982 to examine the Yates thesis. The majority of contributors concluded that although Yates's claim for Hermeticism as the decisive force in paving the way for the Scientific Revolution was exaggerated, her basic insight into the manifold connections between esotericism and science in the early modern period is beyond dispute. The most trenchant deconstruction of Hermeticism in this volume appears in Copenhaver's essay. While not disputing the impact that magic had in fostering the notions of manipulation and control so central to the development of modern science, Copenhaver persuasively argues that the real source of magic theory was not Hermeticism but Neoplatonism. He therefore cautions those scholars who dismiss magic as unimportant in the formation of modern science because they reject the importance of Hermeticism and suggests that they take into account all sources of Renaissance magic—Neoplatonism, Plato, Aristotle, Galen, Aquinas, and Albertus Magnus.[28]

Thus while Hermeticism was demoted from the preeminent position Yates ascribed to it, the participants in the Folger seminar agreed that esoteric currents of thought fostered the more optimistic view of human nature that was an essential prerequisite to the Scientific Revolution. This, too, has become a point of contention, however, for those claiming that patristic theology laid the foundation for such optimism.[29] Augustine has been singled out in this regard; but to find a positive view of human nature in Augustine, Humanists would have had to read his works very selectively, virtually ignoring the later writings.[30] In his biography of

Augustine, Peter Brown evokes the extreme pessimism of these later works:

> God had plainly allowed the human race to be swept away by His wrath. . . . The human race is "the Devil's fruit-tree, his own property, from which he may pick his "fruit." . . . The demons . . . have been enrolled as the unwitting agents of a superior justice; but it is they who are seen as active, and man as merely passive.[31]

Even if one agrees that Augustine allowed a measure of freedom to man, this freedom appears utterly different from the kind of freedom envisioned by the Florentine Platonists. In Augustine, as in all orthodox patristic and Christian writings, there is the inescapable fact that man must rely on grace for salvation.[32] This idea is virtually eliminated in the Florentine Platonists and the numerous sixteenth and seventeenth century proponents of esotericism. Pico della Mirandola's description of man's ability to become whatever he wishes, even to the point of divinity, is a key text for this view of human potential. As we have seen, Pico places absolutely no limitations on man. He has only to "aspire" and to "will," both key words in the *Oration*, and he will find himself "inferior to . . . [the angels] in nothing. "[F]or if we will . . . ," says Pico, "we can." Pico's conviction that humans have the ability to control their own destiny without any external help appears in Ficino as well. Ficino describes the human striving to "become God" as entirely "natural," in the same way that flight is to birds. Men do not require divine assistance to become perfect in this world.[33] Like Pico, Ficino quotes the passage in the Hermetic text describing man's ability to become anything he wishes and he too subscribes to the Hermetic view of man as "a great miracle."[34] This was the view of Agrippa as well: "for by how much the more we have relinquished the animals and the human life, by so much the more we live like angels, and God, to which being conjoined and brought into a better condition, we have power over all things, ruling over all."[35]

One might argue that even this exalted vision of man's potential remains essentially Christian because Ficino is careful to say that it is God who invests the power of perfectibility in human beings. But the very fact that this power is innate obviates the need for either Christ or the Catholic Church, and in taking such a radical position Ficino, Agrippa, along with Hermeticists and alchemists were anything but orthodox. The sense of sin so remarkable in Augustine is absent in these self-professed Christians, and in this they revealed their Neoplatonism far more than their Christianity.[36] The idea that an individual has the ability to take charge of his own salvation as well as the world's is inherently un-Christian, and yet it forms the basis of the modern idea of progress. The future envisioned by the notion of progress is a product of immanent development and does not occur as a result of transcendental, divine intervention (*deus ex machina*).[37]

The positive evaluation of the active life one finds in Ficino, Pico, Agrippa, Della Porta, the Rosicrucians, and seventeenth-century natural philosophers and millenarians is qualitatively different from anything in patristic, medieval, or early humanist writings. Ficino offers a vision of man as a Gnostic savior, something that is inconceivable in Augustinian theology. One of the reasons man is so potentially powerful is that he is a master of the mechanical arts. During the Middle Ages the mechanical arts were tainted by their association with manual labor and concern with worldly ends. The widespread etymology deriving the term from the Greek word for adultery (*moicheia*)—on the grounds that the mechanical arts trick and deceive—mirrored the widespread view that art was vastly inferior to nature.[38] Ficino's positive view of man's ability to use his talent, knowledge, and skill in the mechanical arts to shape and change his environment marks a radical rejection of medieval attitudes.[39]

Such a conclusion is supported by the condemnation of curiosity that runs like a leitmotif through the medieval period. Carlo Ginzburg emphasizes the effect that Jerome's mistranslation of Romans 11:20 had in turning curiosity into a sin. Jerome translated Paul's injunction "be not high-minded" as "noli altum sapere," which was consistently interpreted to mean "do not seek to know high things."[40] No one was more critical of human curiosity than Augustine, who defined it as one of the three forms of the vice of concupiscence: lust of the flesh, lust of the eyes, and worldly ambition. Curiosity was "lust of the eyes," and Augustine links it directly to both the sin of pride and the Fall.[41] Furthermore, curiosity was the source of heresy and the black arts.[42] The association of curiosity with sin, pride, heresy, and magic became conventional in medieval thought and is enshrined in the Faust Legend. It has been argued that this legend was a product of the Reformation and particularly of Lutheran thought, but there is ample evidence to show that it fit in with the medieval Church's condemnation of curiosity.[43] Faust was only one of many figures damned for their curiosity. Icarus, Daedalus, Prometheus, and

Proteus were others denounced either for aiming too high or, as in the case of Proteus, for adapting too easily to changing circumstances.[44] The Hermetic image of man as marvelously protean *and* curious *and* Faustian, the image that exhilarated Ficino, Pico, and natural magicians in general, did not fit well with the medieval view of the world as an ideally static hierarchy, where individuals assumed, and were expected to maintain, the occupations and roles into which they were born.[45] This image did fit in well, however, with natural philosophers and aspiring individuals in the early modern period with the emergence of the notion of progress and the so-called "enterprise society."[46] The accumulated evidence would therefore suggest that the Middle Ages were not fertile ground for the optimistic evaluation of human potential and power that leaps forth from the pages of alchemists, Kabbalists, magicians, Pansophists, and Rosicrucians.

There were, however, two medieval sources for such ideas: alchemy and black magic. The idea that medieval alchemical writings are the place to look for a validation of technology and human potential has also been proposed by a number of scholars.[47] Both alchemy and magic were identified with Hermes Trismegistus. Alchemy was known as the "Hermetic Art," and Hermes Trismegistus was its reputed founder. Hermes is also mentioned with respect in the notorious magical text *Picatrix*. Furthermore, although the Hermetic view of man as a potentially divine Gnostic savior was an aspect of alchemy from its inception, this idea was transformed and strengthened from the fifteenth to the seventeenth centuries, and it was on this later tradition that natural philosophers and defenders of natural magic like Agrippa and John Dee drew.[48]

Thus alchemy may indeed provide a key for the transformation of the medieval into the modern world. Alchemy valued transmutation, in other words, change and evolution. It was oriented toward this world and the improvement of it. It proclaimed the power and liberty of the individual to determine his own fate without the mediation of any institution or authority.[49] In an age of bitter sectarian warfare, alchemy provided a refuge for those who clung to the Renaissance ideal of a universal philosophy or *prisca theologia* that would unite men in a common quest to restore the world to its prelapsarian perfection. Alchemists were essentially a fifth column within every Christian denomination; they carried forward the optimistic ideals of Renaissance Platonists into the age of the Enlightenment. As we have seen, Boyle, Locke, Leibniz, and Newton all studied alchemy.[50]

Drawing on the work of Betty Jo Dobbs and other scholars, Bruce Moran gives pride of place to alchemy for jumpstarting the Scientific Revolution. He argues that for all its esoteric trappings alchemy was based on procedures that could be replicated and taught. It was an intensely empirical discipline, dedicated to the kind of experimental research that fed into branches of modern science, and it succeeded in producing many items from medicines to industrial products that improved the quality and enjoyment of life. Furthermore, alchemy was a key factor in promoting the idea that human beings had the intelligence and ability to improve the world. Alchemists tipped the scales in favor of art over nature and in so doing fostered the belief in progress that became the hallmark of modern science. Moran contends that alchemists were concerned with the same problem as physicists and astronomers about what sort of "active principles" existed to account for the attractions and affinities between bodies. For all these reasons alchemy deserves a recognized place in the history of the Scientific Revolution.

Moran stresses the fact that science is not just a "cognitive realm," but an "existential one." He follows recent scholars out of libraries and into new places—courts, artisan workshops, pharmacies, botanical gardens, local scientific societies, and even kitchens.[51] Alchemy was practiced in all these places, and while gold-making was a constant feature, there were other important branches concerned with the distillation of medicines, metallurgy, the investigation of chemical processes, and the manufacture of useful products such as dyes, inks, artificial gems, gums, resins, acids, and cosmetics, to name a few. What bound all these activities together was their common attempt to transform physical substances into something better than what they naturally were. In this sense, alchemists were like bakers who transformed flour into bread or vintners who turned grapes into wine; they improved on nature. Although alchemical texts could be obscure, for the most part they describe procedures in a manner that could be decoded and in this way added significantly to the accumulated store of chemical knowledge. Distillation techniques were primarily used for making medicines, and while these may not have always been effective, the distillation of substances like Benedictine (and other sorts of *acqua vitae*) helped (and still helps) to dull the pain. The discovery of new products is what makes alchemy "such an important feature of the Scientific Revolution."[52] Books like Agricola's *De re metallica* (1556) and Biringuccio's *Concerning the Making of Things by Fire* (1540) presented readers with the technical details involved in mining and metallurgy and a host of procedures to make such things as steel and glass, distill mercury, and purify saltpeter, salt, and vitriol through crystallization. The techniques presented in these works contributed to economic developments in the early modern period and explain why alchemists and artisans were welcome at many European courts. Princes were eager to enhance their finances and reputations, and they were willing to pay and, in some cases, even kidnap alchemists to do just this.[53] In addition to technical manuals, there was a genre of literature know as "Books of Secrets," which presented readers with recipes for removing stains, soldering, etching, coloring metals, making artificial gems, dyes, shoe polish, and even cream to whiten the skin and remove wrinkles. One of the most popular of these was *The Secret of Lady Isabella Cortese* (1561). Whether this was actually written by a woman or not, historians are slowly uncovering the role that women played in alchemy. Because their domain was the kitchen and caring for the health of family members, women's involvement in alchemy is understandable. Books of Secrets marked the beginning of the kind of scientific and technical works that are so prevalent in the "How-to" sections of bookstores to this day. As individuals searched for secrets, they discovered new products and new ideas and gained the experience, confidence, and ability to appreciate novelty and challenge accepted wisdom, both characteristics of the Scientific Revolution as well as modern science.

The contribution made by alchemy to the emergence of modern science underlines the error of seeing the Scientific Revolution as the triumph of good science over bad pseudoscience. In Moran's view, the Scientific Revolution simply was the mix of conflicting ideas and ideologies that existed in the early modern period. This "concoction" of incompatibilities stimulated the discoveries associated with the emergence of modern science: "the Scientific Revolution . . . has much to do with the presence of impurities of various sorts—the sometimes inharmonious intellectual and social mixture of learned and artisan, of occult, spiritual, and mechanical. This is the concoction that woke things up and produced a cultural reaction."[54] This still leaves us with the question of why people wanted to smooth out inconsistencies and what made them so interested in new products and inventions, which takes us back to the activism and the more positive vision of human nature fostered by religious and esoteric currents of thought.

If there is one factor that defines the modern world above all others it is science, and, as we have seen, historians have endlessly debated precisely what conditions and what factors contrived to make modern science develop in one and only one place, Western Europe. Yates's contention that Hermeticism provided the outlook and optimism conducive to the development of modern science is only one of several possible answers that have been given. The explanation most prevalent before hers was Protestantism. Robert Merton's essay was the first in a long line of books and articles that argued for a Protestant and particularly a "Puritan" impetus in the development of science.[55] Merton was well aware that science could and did thrive in a Catholic environment. After all, Copernicus, Kepler, Galileo, Mersenne, Pascal, and Descartes lived and worked in Catholic countries, and the discoveries made by Catholic and continental mathematicians paved the way for the extraordinarily accomplishments of Newton.[56] Merton was therefore careful to stipulate that he was not claiming specific Protestant doctrines promoted or impeded science, simply that certain Protestant "sentiments" and "dominant values" were especially conducive to science in seventeenth-century England. The problem is that he never clearly spelled out exactly what these sentiments and values were in any more precise way than by suggesting that Protestants showed a penchant for a utilitarian and experimental approach to science. These ideas have, however, been taken up and refined by later scholars, all of whom are and should be greatly indebted to Merton for being one of the first scholars to recognize that religion and science were not necessarily antagonistic but often mutually supportive and for appreciating the indirect role that social and economic factors have in validating science and the interests scientists pursue.

On the basis of Peter Harrison's work one might venture to say that Merton was too cautious, that it really was a matter of doctrine when it came to support for science among early modern Protestants, the doctrine in question being that of original sin or, more precisely, the rejection or modification of the doctrine of original sin. However, this rejection came primarily from esoteric and millenarian thinking outside the bounds of orthodox denominational views, whether Protestant or Catholic. But there is also the fact of biblical literalism, which began with Protestants and which, as Harrison persuasively argues, was a key factor in generating the kinds of questions that encouraged the pursuit of science.[57] Finally, one cannot overlook the effects of the Inquisition and Index. Recent scholarship has tended to mitigate the consequences of both. While the Catholic Church was not the draconian institution imagined by Protestants and earlier historians of science who were largely Protestant, the dampening effect of the Council of Trent and Galileo's trial on Catholic scientists cannot be ignored. The fact that Galileo ended up under house arrest, while John Wilkins, an advocate of Copernicanism, became an Anglican Bishop provides a telling example of the different fates in store for natural philosophers living under different religious jurisdictions in early modern Europe.[58]

The difficulty of asserting a direct link between theology and science had the effect of turning historians in the opposite direction, to the point that some claimed religious moderates and religious moderation laid the foundation for both the Scientific Revolution and the Enlightenment.[59] The idea that toleration and moderation were what mattered was first suggested by Hugh Trevor-Roper when he proposed that one look to Erasmus as the real progenitor of the scientific spirit with its combination of skepticism and a belief in the legitimacy of rational, critical inquiry.[60] But Erasmus was no Newton, nor even a Francis Bacon. While he did advocate those aspects of mind essential to critical and methodical investigations of all kinds, there is no hint in his writing that man can and should transform his world as well as himself. Erasmus lacks the passion and commitment to reform so characteristic of many early modern philosophers and scientists. It is virtually impossible to see Erasmus as the source for the kind of prophetic and millennial rhetoric one constantly meets in such thinkers as Bacon, Comenius, Hartlib, and the host of natural philosophers who did so much to popularize the view that science was the answer to human pain, discomfort, and distress. Even Boyle exhibits hints of enthusiasm when he argues that scientists are the true priests. In short, Erasmus's measured view of human potential lacks the confidence and optimism that marks the work of seventeenth- and eighteenth-century scientists. While this optimism and confidence appears most vividly in the millenarian schemes discussed in the previous chapter, it emerges in another area as well, in the innumerable discourses on method that are such a feature of early modern thought.

Although Descartes was deeply affected by the skeptical crisis generated by the Reformation, this did not stop him from believing that men could discover the truth if they employed their reason according to the dictates of his prescribed method. His method may not inspire much confidence today, but the fact that it inspired him is indicative of his positive assessment of the potential of the human mind. From a modern vantage point it is hard to understand the rapture that greeted the many and various methodologies proposed in the early modern period or to comprehend exactly why people accepted at face value the glorious prospects they promised. All the major works dealing with grammar and teaching in the early modern period contain what we can only dismiss as wildly exaggerated claims about the efficacy of method. The scholar and linguist Franciscus Sanctius (1523–1600), for example, claims he can teach Latin in eight months, Greek in a mere twenty days, and astronomy in little over a week.[61] Abbot Trithemius, considered a demonic magician by a number of his contemporaries, makes even greater pedagogical claims. He tells a friend that by means of his stenographical art he can "teach an uneducated man in only two hours, though he be knowledgeable only in his mother tongue and have never known a word of Latin, to write, read, and comprehend whatever he is meditating upon in Latin, and this with sufficient elegance and skill that those who see his writing will praise his words and understand his Latin composition."[62] Leibniz asserts that he discovered everything from a calculating machine to a submarine as a result of his new method or "Combinary Art." Such astonishing claims to such astonishing results achieved solely through the application of some prized method had been a standard aspect of memory treatises for the previous two centuries.[63]

The growing faith in methodology provides an example of the transformation of millenarian and esoteric thought into the secular faith in progress. This faith remains very much alive today as the availability of an enormous variety of psychotherapies reveals. However unrealistic, the belief in the existence of a method capable of solving every problem is immensely satisfying emotionally. Only in our psychologically oriented era have historians considered this dimension of the passion for method in the early modern period:

> At its most basic, method functions as a guarantee of order, if in no other way than psychologically. If we can methodize, follow a strict set of rules, consciously proceed by an institutionalized (impersonal) set of directions, we have some sense of triumphing over, or at least controlling, chaos.[64]

In a very real sense during the sixteenth and seventeenth centuries method promised some of the things religion had in earlier centuries. It is therefore interesting that method, as in "Methodism," should be attached to religion in the following century. Even Descartes's search for the perfect method, taken by previous generations as an illustration of his rational approach to knowledge, is now interpreted quite differently, as evidence of his compelling need to overcome a profound skeptical crisis that left him in psychological turmoil.[65]

The belief that reason combined with method provided a virtually infallible formula for scientific discovery and scientific progress became the accepted wisdom of the Enlightenment and subsequent centuries, providing the basis for the notion of progress. The origin of this new and positive attitude toward the world and mankind lie in sources that were in essential ways antithetical to orthodox Christianity. Esotericists and their followers undoubtedly believed that Neoplatonism, Hermeticism, the Kabbalah, alchemy, and all the other forms of esotericism they valued were compatible with Christianity, but in significant ways they were not. Each of these philosophies preached the doctrine of perfectionism and in so doing undermined the Christian emphasis on the unique role of Christ as the *only* and *transcendent* source of salvation. In a very real sense the alchemist was Christ, the Neoplatonist was a powerful magus able to draw divine power into himself, and the Kabbalist was a link between God and creatures responsible for maintaining the universe. Instead of locating the source of this confidence in the magical and occult traditions of the Renaissance one might perhaps go a step further and find it in the revival of Pelagianism during the Renaissance.[66]

The revival of Pelagianism in the Renaissance is surely of great importance in the shaping of the later scientific mentality. However, the reason why Pelagianism came back into favor during the Renaissance was because of the rediscovery of texts embodying Greek and Roman scientific, philosophical, esoteric, and magical traditions. These texts precipitated a confrontation between philosophy and theology, on the one hand, and science and religion, on the other, and they encouraged a reevaluation of the goal and potential of human reason. A more positive view of both nature and human nature emerged from this confrontation. Among these rediscovered texts was, of course, the *Corpus Hermeticum.* The rediscovery of a wide variety of texts—and these include alchemical, Neoplatonic, and Kabbalistic works—led to the resurgence of the Pelagian heresy with its belief in the ability of man to determine his own destiny as well as that of the world. As Trevor-Roper suggested many years ago, the origins of the

Enlightenment are to be found in heresy or, if not, outright heresy, at least heterodoxy:

> Thus, when we look into the religious origins of the Enlightenment we do not discover them in any one Church or Sect. They are to be found in both Churches and in several sects. What is common to the men who express such ideas is that all of them are, in some sense, heretical. That is, they either belong to dissident groups within their Churches or are themselves regarded as unorthodox. The orthodox Churches—Catholic, Lutheran, Anglican, Calvinist—look askance at them.[67]

The last two chapters have shown that while scholars do not agree on precisely what factors laid the foundation for the Scientific Revolution, during the past forty years it has become increasingly apparent that religious and esoteric currents of thought must be taken into consideration. Historians have come to realize that there is no way to separate the sheep of science from the goats of religion and magic. Up to at least the mid-eighteenth century, and in some cases well beyond, they all came in a single flock. It is incumbent on us to remove our modernist spectacles and see how subjects, disciplines, and categories that seem so distinct now could have coexisted, overlapped, and changed their meanings over time. To illustrate the need to do this as well as to show what scholars have already accomplished, the next chapter offers a case history, revealing the complex strands of religion, magic, and science in the work of J. B. van Helmont, Robert Boyle, and Isaac Newton.

CHAPTER 9

A Test Case

Postmodernismis well known for denying the existence of absolute values, limiting the scope of human rationality, rejecting metanarratives, and insisting on the social construction of truth. In the history of science, postmodern scholarship has had the effect of discrediting the grand narrative prevailing until the 1960s that described the Scientific Revolution in "Whiggish" terms as a succession of great discoveries made by great men, who abandoned magic and religion for science proper.[1] As we have seen, during the past fifty years astronomy, physics, and mathematics have been demoted from their primary position as Aristotelianism was brought back into the picture and noncanonical subjects such as alchemy, magic, Neoplatonism, Kabbalah, natural history, and antiquarianism were added. The recognition of esotericism as a distinct and worthwhile field of study has had a major impact on these developments, and, as a consequence, a new picture of the Scientific Revolution as well as the Enlightenment has emerged, in which religion is recognized for the role it continued to play.

One clear consequence of this ferment in the history of science is a new appreciation of the contributions made by alchemy to the Scientific Revolution. Drawing on the work of many scholars, Bruce Moran has argued that alchemy was not irrational but dedicated to the kind of experimental research characteristic of some branches of modern science and involving laboratory procedures that could be replicated and taught. William Newman and Lawrence Principle take the same position. They provide conclusive evidence that there was no break between alchemy and chemistry. Alchemy did not lack, as many earlier historians have claimed, important qualities of "modern" laboratory practice such as quantification, theory-guided practice, practice-informed theory, and reproducibility.[2] The foundation for the eighteenth-century chemical revolution associated with Lavoisier lay in the quantitative and experimental work of medieval and early modern alchemists, or "chymists," as Newman and Principe call these forerunners of modern chemists.[3] While not enough is yet known to write a complete history of alchemy in the early modern period, what can be said is that there was an explosion of interest in the subject marked by the publication of an enormous number of alchemical texts. A relatively marginal pursuit in the Middle Ages, alchemy became a mainstream occupation, influencing scientists and philosophers as well as artists, poets, mystics, Freemasons, theosophists, esotericists, and spiritualists of all sorts. The fact that many of the greatest scientists and philosophers of the early-modern period such as Robert Boyle, John Locke, Isaac Newton, and Gottfried Wilhelm Leibniz performed alchemical experiments reveals alchemy as a field of importance for understanding the emergence of modern science and the transition from the medieval to the modern world.

The traditional view that chemistry experienced a "postponed scientific revolution" until the eighteenth century when Antoine-Laurent Lavoisier (1743–94) overthrew the phlogiston theory has been effectively laid to rest by historians who stress the evolutionary nature of developments in chemistry.[4] This is not to deny Lavoisier's own contributions, but it does reveal how Lavoisier drew on the experimental findings of his predecessors, particularly J. B. van Helmont (1579–1644), whose own work brought together key quantitative aspects of medieval alchemy with the theory and practice of Paracelsus. Paracelsus (1483–1546) is a key figure in the transition from alchemy to chemistry. He and his followers broadened the scope of alchemy to encompass investigations of the composition, structure, properties, and reactions of matter. They were especially interested in chemical medicines. As experimenters, textbook writers, and systematizers, Paracelsians had a significant impact on medicine, pharmacology, and chemistry. Their interest in medicine led to the establishment of a school of "iatrochemisty," or medical chemistry, that attacked the foundations of Aristotelian natural philosophy and Galenic medicine, attracting many adherents and becoming an accepted part of the curriculum of medical schools during the seventeenth century.[5] As perhaps the most famous (or infamous, according to his detractors) follower of Paracelsus, Jan Baptista van Helmont took a prominent part in the attack on Aristotle and Galen, and his ideas influenced the work of later chemists such as Robert Boyle and Isaac Newton, both consummate students of alchemy who owned and read van Helmont's most important works. Van Helmont, like Paracelsus, provides an instructive example of the importance of religious and esoteric thinking in the early modern period, revealing that both were integral to the science of the time.

Van Helmont's published work offers a mixed bag of metaphysical speculations and accurate laboratory observations. It was the work of a scientist dedicated to experimentation and the exact analysis of chemical and biological phenomena, but it was also the work of a man whose religious convictions profoundly motivated the course of his research. Walter Pagel has shown how entwined religion and science were in the elder van Helmont's thought, a situation that was not unique to him. The ideal of separating religion and science proposed by Galileo in his letter to The Grand Duchess Christina (and supposedly practiced by the members of England's Royal Society) was just that, an ideal. In reality, seventeenth-century natural philosophers were deeply influenced by their religious beliefs, as we have seen. One of the defining aspects of van Helmont's philosophy was his rejection of what he described as "pagan" books and "pagan" learning and his emphasis on the discovery of knowledge through experiment, revelation, and the introspection of the innate wisdom hidden within the human soul. This was also one of the crucial points in the acrimonious debate between Paracelsians, on the one hand, and Galenists and Aristotelians, on the other, a debate in which van Helmont played a conspicuous part. Van Helmont had nothing positive to say about his Aristotelian and Galenic opponents. He attacked and ridiculed them on an intellectual, moral, and institutional level. Like his mentor Paracelsus, he was neither modest in presenting his opinions nor restrained in his language. His condescending and at times rebarbative attitude so infuriated the scientific and medical establishment that an acrimonious battle of books ensued that did not end well for van Helmont.

One of the major bones of contention between Paracelsians and Galenists was over the cause and cure of diseases.[6] The Galenists followed Aristotle in conceiving of matter as consisting of the four qualities—hot, cold, moist, and dry—which made up the four elements of earth, air, water, and fire. These four elements corresponded to the four humors—blood (air), phlegm (water), yellow bile or choler (fire), and black bile or black choler, more commonly known as melancholy (earth). Health required a balance of these four humors, while disease was caused by their imbalance. There was therefore no concept of *diseases* among Galenists, only of *disease*, and this was not a positive entity in its own right, but a negative state resulting from the imbalance of humors. Van Helmont rejected the entire notion of the four elements upon which Aristotelian philosophy and Galenic medicine were based. In this he followed Paracelsus; but whereas Paracelsus had replaced the four elements with his three principles of salt, sulphur, and mercury, van Helmont rejected these three principles as well. He argued instead that the basic constituents of bodies were water and *semina*, the latter being immaterial "seeds" or seminal principles that turned water into uniquely individual entities. Van Helmont rejected the four elements and the three principles on experimental grounds. He argued that Aristotelians, Galenists, and Paracelsians were all mistaken in thinking that fire, or combustion, separated bodies into their constituent parts. Instead of reducing bodies to their elements, combustion actually produced the substances that were then observed. This observation provided seventeenth-century chemists like Robert Boyle with an important conceptual tool for reassessing the status of those substances commonly considered to be ultimate principles.

According to Galenists, one of the principal causes of disease was catarrh. Catarrh was a cold, watery humor, or phlegm, produced in three stages: first, vapors ascended from the hot stomach to the cold brain; the brain acted much like a lid on a pot of boiling water on which the steam condenses. Once condensed, these vapors trickled down into the body, causing all manner of maladies in the nose, lungs, joints, bones, and organs. Catarrh was essentially a kind of poisonous excrement that destroyed the balance of humors. Given their diagnosis of disease, Galenic physicians resorted to therapeutic treatments aimed at restoring the balance of humors. This could be done positively by prescribing certain foods. Because food also consisted of humors, the goal of the physician was to prescribe those that would augment a patient's deficient humors or decrease predominant ones. When Sir Andrew comments in *Twelfth Night* (I, iii), "I am a great eater of beefe, and I believe that does harme to my wit," he is thinking like a Galenist. Other forms of Galenic therapy were purging and bleeding, which were designed to draw off excessive humors.

Paracelsians had nothing but scorn for these core concepts of Galenic medicine. In a bit of doggerel van Helmont ridiculed Galenic diagnostic techniques, which mostly consisted of examining a patient's urine and feces for evidence of an imbalance of humors: "Excrementitious dung and Urine piss/are of Physitians, the chief dainty-dish."[7] Not only did the Paracelsians and Helmontians disparage the medical practices of Galenic physicians, but they attributed them to the worst possible motives: "Our wish is that . . . they [Galenic physicians] would . . . no more attempt . . . to dreigne the hopes, bodies, veins, strength and purses of the sick."[8] Denying the existence of humors, van Helmont went so far as to attribute their fabrication to the devil himself, an assertion bound to infuriate the medical establishment: "for truly Humours are destructive Ignorances, sluggishnesses, introduced by the Father of lies, and celebrated by the loose credulity of his followers."[9] If humors did not exist, neither did catarrh, a fact that van Helmont drove home in a treatise with the abrasive title, *Deliramenta Catarrhi: or, the Incongruities, Impossibilities and Absurdities couched under the Vulgar Opinion of Defluxions.* In this tract he ridicules the whole concept of vapors on the grounds that there simply are no spaces in the head to accommodate the purported trickle-down effect. Because both humors and catarrhs are Galenic fictions, cures based on diet, purging, and bleeding are worthless to the patient; they merely serve to line the pockets of avaricious physicians.[10] Van Helmont's concern for the poor was a hallmark of Paracelsian medicine, which was predicated on a radical equalitarianism. Just as Luther argued for the priesthood of all believers, Paracelsus undermined the rationale for class distinctions with his claim that all men had equal access to divine wisdom inasmuch as every man was a microcosm and possessed within himself knowledge of the greater world. Paracelsus called for an end to the social distinctions separating physicians from apothecaries and surgeons. In the same way that Luther considered each individual's "calling" or occupation equally valid in the eyes of God, Paracelsus believed that all men took part in the cosmic drama of world redemption.[11]

The point that van Helmont and all Paracelsians stress in their attacks on Galenic medicine is that humors and catarrhs are not the cause of disease, but the pathological products of diseased bodies. There is no point in attempting to remove these products unless the cause is dealt with. In van Helmont's view, the Galenists were like someone who treats a wounded finger without removing the thorn that caused the wound. The claim that diseases are caused by specific entities that attack specific organs marks a positive contribution to medical theory. For Paracelsians there are many different diseases, each with its own life cycle and specific pathological effects. This concept of diseases as active, independent entities led Paracelsus and van Helmont to several strikingly modern ideas: (1) diseases can lie dormant in the body until suitable conditions arise to activate them—this suggests something like an incubation period;[12] and (2) certain diseases are more likely to attack certain people in certain places. Paracelsus was the first person to write a complete monograph on an occupational disease. In his *Von der Bergsucht und anderen Bergkrankeiten* he describes the pathological conditions affecting the lungs and organs of miners as a result of their inhalation of mercury vapor.[13]

When it came to explaining how a specific disease attaches itself to and affects a particular organ, both Paracelsus and van Helmont argued that diseases cannot affect the body in a straightforward, physical way. Diseases are products of the imagination. Van Helmont aligned himself with the vitalists, claiming that what we call matter only exists as certain specific entities organized by the ideas within them.[14] He criticized the ancient atomists and the scholastic and modern philosophers influenced by them for believing that nature was formed from inert matter and consequently amenable to mathematical and mechanistic explanations. This was, of course, a hotly debated issue in the seventeenth century. Van Helmont did not, however, reject mathematics wholesale, only the scholastic "mathematics" taught to him during his short career as a university student at Louvain, where a sort of speculative mathematics, with no basis in actual observation or laboratory experiment, was applied to humors and the course of fevers and diseases. Although van Helmont "felt a profound repugnance" for this kind of "mathematics," he believed that university students should be taught the kind of practical mathematics used in navigation, surveying, and mapmaking. He also favored chemical analysis in terms of careful quantitative measurement.[15] J. R. Partington pointed out years ago that van Helmont clearly expressed the law of the conservation of matter.[16] Even more important was his idea that in every chemical reaction weight, along with matter, was always conserved, an idea that provided the basis for the modern idea of "mass balance."[17]

Although van Helmont would have been the last to admit it, or perhaps even realize it, he was profoundly influenced by the Aristotelian doctrine that "form" is inherent in matter and inseparable from it. Like Aristotle, he believed that there is a principle within each material body that determines its life history.[18] Borrowing from the vocabulary of

Paracelsus, van Helmont called this life principle the *archeus* (Greek for a beginning, origin, first cause). Complex organisms like animals or humans contain many *archei*, which control various organs and limbs. Like soldiers under a general, these are controlled by a supreme *archeus*. Van Helmont's scientific thought is essentially an elaboration of the role these nonmaterial forces or *archei* play in nature. In arguing that material bodies can have no effect whatsoever on each other, van Helmont made use of ancient ideas about the power of the imagination and the role of spirits. These ideas continued to be influential well into the nineteenth century among, for example, Mesmerists, Spiritualists, and Parapsychologists, and they exist in a much modified form today with the recognition of the psychosomatic causes of illness.

The example van Helmont gives to explain the interaction between *archei* is of a man who thrusts his hand into a pail of pitch. He denies that the man's hand becomes black because the material particles of the pitch stick to it, but attributes the blackness to the effect the *archeus* in the pitch has on the *archeus* in the hand.[19] Exactly the same process occurs in the case of disease. The alienation of the *archeus* of a body or organ through the influence of an "idea morbosa" is the true cause of illness, not imbalanced humors or fluxating catarrhs:

> A disease ... is a certain Being, bred after ... a certain hurtfull strange power hath violated the vital Beginning, and hath stirred up the Archeus into indignation, fury, Fear, etc. ... the anguish and troubles of which perturbations do by imagining, stir up an idea co-like ... themselves, and a due Image: Indeed that Image is readily stamped, expressed and sealed in the Archeus, and being cloathed with him, a disease doth presently enter on the Stage.[20]

As specific entities, diseases have their own life rhythms, and it is the job of the physician to study and cure these through careful observation. Unlike Galenic physicians, who could make diagnoses from urine samples in the comfort of their own homes, Paracelsians had to remain by the bedside. With their new concept of the etiology of disease came new therapies. Because diseases are caused by morbid ideas, remedies must be designed to stir up healthful ideas in the disturbed *archeus*.[21] Just as diseases are specific entities, so are remedies, which must therefore be carefully prescribed. Paracelsus and his followers are primarily remembered for their advocacy of chemical, as opposed to organic, medicines. Such chemical remedies are necessary because Paracelsians believed that diseases cause chemical changes in the body.

Having denied that materialistic explanations account for the activities of a healthy or diseased body, van Helmont had to explain how nonmaterial ideas could be the cause of physical changes. This was a troubling question that exercised a great many theologians, demonologists, and natural philosophers writing before and after him. Van Helmont's solution lay in identifying various spirits of different degrees of subtlety, which the soul, or *archeus*, employed to direct the body toward some specific end, a common solution with ancient roots.[22] For example, the "vital spirit" was the medium through which a thought in the soul was passed to the body: "The *vitall Spirit* in the throne of flesh and blood, that is the outward man, sits viceroy to the *Soule*, and acts by her commission."[23] The vital spirit was not the first instrument of the soul, however. There was another more rarefied spirit that emanated directly from the soul itself:

> The *Soule* therefore, by essense wholly *spirituall*, could by no means move, inform, and actuate the *vitall spirit* (which truely carries something of *corporeity* and bulk) much lesse excite and give locomotion to flesh and blood; unlesse some *naturall*, yet *magicall* and *spirituall* power inhaerent in the *soule*, did streame down from the *soule*, as from the first motor, upon the spirit, and so descend to the body.[24]

Every created thing possessed spirits to some degree, even objects like pitch, otherwise there could be no interaction between bodies. Man, however, as the image of God and apex of creation, has been blessed with the most subtle and potent spirit of all:

> Hitherto Have I suspended the revealment of a grand mystery: namely, to bring it home to the hand of reason, that in man there sits enthroned a *noble energy*, whereby he is endowed with a capacity to act *extra se*, without and beyond the narrow territories of himself, only *per nutum*, by his single beck, and by the natural magick of his *phansie*, and to transmit a subtil and invisible *virtue*, a certain influence, that doth afterward subsist and perservere *per se*, and operate, upon an object removed at very large distances.[25]

With the same kind of exhilaration and wonder Pico della Mirandola had expressed in his *Oration on the Dignity of Man*, van Helmont here describes man as a magus. He qualifies this statement, however. In origin man is a magus, but sin has enfeebled his powers. They may be restored but only through the gift of divine illumination.[26] The idea that spirits exist that provide the soul with the means to influence its own as well as other bodies was commonly accepted at the time van Helmont wrote.[27]

As Alexandre Koyré has said, "we are all Cartesians, whether we want to be or not."[28] It is therefore difficult for us to grasp the concept of spiritualized matter that crops up continually not only in the writings of Paracelsians but also in many other mystics, poets, and natural philosophers. The Plotinian doctrine of creation as a process of emanation from a single spiritual entity entailed the notion of continuity between spirit and matter.[29] Later writers, particularly Neoplatonists like Ficino, Telesio, and Campanella, tended to make the spirit a bridge between the two and eliminate souls altogether, a practice that Paracelsians followed. Van Helmont's discovery of gas encouraged his belief that the operative units in nature consisted of entities that were simultaneously matter and spirit, similar to gas.[30]

When Paracelsians like van Helmont described how natural bodies interact, they draw on numerous theories readily at hand: the Neoplatonic *anima mundi* and astral body, the stoic *pneuma*, and even Galenic medical spirits were all utilized to explain how such interactions take place. Paracelsians believed in the power of the imagination. They followed in the tradition of Avicenna rather than Aquinas, maintaining that the imagination could act immediately on objects even at a great distance. For example, they accepted the idea that the imagination of a mother had the power to affect her offspring. This idea was not discredited until the eighteenth century. As late as 1713 a report appeared in *L'Histoire de L'Academie Royale des Sciences* of a child born with a beef kidney for a head because its mother could not satisfy her craving for kidneys. At an even later date Mary Toft supposedly produced her rabbit offspring, as we have seen.

Paracelsians like van Helmont believed the imagination worked through a kind of sympathetic emanation. This was in no way novel itself, but the way van Helmont elaborated it was. He used the idea of a spirit to explain the occult practices of witchcraft, magic, magnetism, and sympathetic medicine. He even made the grave error of applying analogical arguments drawn from his ideas about spirits to explain in natural terms effects the Catholic Church was bound to insist were supernatural—such as exorcism and the miraculous healing power of relics. (He attributed the effectiveness of both primarily to the power an excited imagination had on the body, through the spirits.[31]) Eventually the Inquisition put him under arrest and questioned him on all these points. The Church was particularly interested in his discussion of the innate power that enabled man to act *per nutum*; the clerics wanted to know how it was awakened, naturally or diabolically? Van Helmont insisted it was a completely natural power. He cited reproduction as an example of the role ideas play in producing corporeal entities, and what could be a more natural act than that?[32] When van Helmont went on to explain the other ways this ostensibly natural power worked, several questionable elements slipped in that the holy fathers deemed diabolical. Eventually he was forced to concede that Satan could stir up the ecstatic power in man and that it was therefore not wholly natural.[33]

The controversy over the "weapon salve" provides another illustration of van Helmont's conviction that spiritual forces are the only active agents in nature. The weapon salve, as its name suggests, was applied to the weapon rather than to the wound itself.[34] The rationale for this procedure lay in Gilbert's work on magnetism. A magnetic attraction was thought to exist between the blood on the weapon and the blood still coursing through the veins of the wounded person. The magnetic attraction ensured that an efflux of spiritual matter communicated between the two bloods, carrying the healing power of the salve with it. It is disquieting to know that this theory lay behind the practice of taking wounded dogs to sea in the hope that communication between the wounded animal and the blood on the weapon left on land would help in determining longitude.[35] In defending the weapon salve van Helmont played into the hands of his Galenic opponents. Intent on silencing him, they brought his views to the attention of the Inquisition, which placed him under house arrest for two years. His interrogators were interested in establishing how van Helmont could defend the action of the weapon salve as "natural" when action at such a distance, if not miraculous, must be diabolical. The Inquisitors were extremely critical of van Helmont's statement that the weapon salve occurred naturally in exactly the same way that miracles did.[36]

The weapon salve controversy is just one of many examples illustrating the belief common to Neoplatonists, Hermeticists, Kabbalists, Paracelsians, and esotericists of all stripes that the world consists of a network of sympathies and antipathies. D. P. Walker has described the ancient roots of this belief, along with its heretical implications.[37] One of the major heretical aspects of this way of thinking is that for Paracelsus and his followers the true physician is a semi-divine being involved in a cosmic process of redemption through the manipulation of these sympathetic forces. The influence of alchemy is obvious in this regard, for the alchemist was thought to have a redemptive and even creative role. Kabbalists credited their actions with the same efficacy. Rejecting the traditional Aristotelian view that nature provides the norm for art, alchemists, Paracelsians, Kabbalists, and magicians believed that through art they could intervene and accelerate natural processes and perfect or redeem matter. This way of thinking was an anathema to many Catholic and Protestant theologians as well as to Aristotelian natural philosophers. The English Aristotelian John Case, for example, criticized Paracelsians for holding such a manifestly false and exalted view of their own powers. He charged them with hubris for setting themselves above nature and for failing to realize that art must follow nature's course.[38]

Galenic physicians were quite naturally incensed by Paracelsus, van Helmont, and their followers, not simply because of what they said but because of the manner in which they said it. Thomas Erastus, for example, was so enraged by Paracelsus's defamatory attack on Galenic medicine that he answered in kind in a long diatribe entitled *Disputationum de medicina nova Phillippi Paracelsi* (1572). Erastus was sufficiently provoked to emulate Paracelsus's habit of making scurrilous *ad hominem* remarks. (The level of vituperation characteristic of debates in the sixteenth and seventeenth centuries make modern debates pale by comparison.) Erastus relished stories about Paracelsus's personal failings. Relying on Johannes Oporinus's (1507–68) account of his brief apprenticeship with Paracelsus, Erastus described the way Paracelsus habitually fell into bed drunk, fully clothed, and armed, often to awaken with a terrifying leap in the air, eyes ablaze and sword drawn. And he repeated with evident relish Oporinus's surmise that Paracelsus was sexually abnormal.

Andreas Libavius was another critic of the claims to illumination made by Paracelsians, and he too was angered by their attack on hallowed medical and academic institutions and conventions. Although he was willing to accept many of the chemical discoveries of the Paracelsians, he was greatly concerned with defending institutionalized learning against what he saw as their nihilistic onslaught. In the first of what became three volumes of letters on chemical subjects addressed to philosophers and physicians throughout Germany, Libavius writes to a young friend to warn him of the potential dangers to his health, social position, and scholarly reputation that would inevitably result from his interest in Paracelsian philosophy. Like Erastus, Libavius emphasizes the impiety and arrogance of Paracelsus and his followers. He is especially critical of their claim to have the God-like ability to perfect nature. Libavius was particularly incensed because they rejected his pedagogical approach. They were not interested in knowing what the most reputable authorities had to say. Their goal was to discover the hidden forms and essences behind the world of appearances. In this they were esotericists to the core and followed in the footsteps of the proponents of the natural magic tradition. But this did not mean that they rejected research and experimentation. In fact, because of their denigration of reason and logic Paracelsians like van Helmont were convinced that knowledge of hidden causes could only come through observation and experiment with the assistance of divine illumination.[39] Van Helmont's stress on illumination and the limits he placed on human reason made his thought particularly congenial to esotericists, religious radicals, and Protestant millenarians.[40] Along with Montaigne and other skeptics, van Helmont denied that man can be defined as a rational creature because other creatures, for example, bees and wolves, are also rational. Reason is, in fact, a product and consequence of the Fall: "[Reason] . . . is properly nothing else, but a wording faculty of discoursing, co-bred with us as mortals from sin."[41] Alluding to the myth of Proteus, van Helmont maintains that the understanding transforms itself into the thing to be understood.[42] The understanding is able to do this because, as we have seen, the thing to be understood is already in the intellect: "Our Soul understanding it self, doth after a sort, understand all other things, because all other things, are in an intellectual manner in the Soul, as in the Image of God."[43] The implied optimism of this statement was tempered, however, by the fact of original sin. Living after the Reformation, van Helmont's view of man was colored both by the growth of skepticism and by the post-Reformation emphasis on man's fallen nature and need for grace. There is a basic difference between his philosophy and that of the Renaissance Neoplatonists in the repeated emphasis he places on man's Fall and consequent need for divine illumination. Man must offer his whole self to God; he must "will, act, and suffer anything, with a total amorous offering up of the heart, soul, and strength into the obedience of the Divine Will."[44] Only then will he gain knowledge: "whether the understanding be transformed, or whether it doth transform it self into the Image of the thing understood, surely it had need of help from God."[45] The emphasis van Helmont places on the "heart" and on divine illumination as sources of knowledge became the defining characteristic of the new forms of spiritualism that proliferated during the sixteenth and seventeenth centuries. This is another indication of how closely entwined religion and science were in the early modern period. But for all the places in van Helmont's published work where he denigrates reason in favor of revelation, it must be understood that he believed divine revelation would only come as a result of continual and careful laboratory work. Flashes of inspiration, even dreams that provided insight into chemical reactions, may come from God, but only as a result of assiduous experimentation.[46] Although a Catholic, van Helmont reveals a mindset here similar in many ways to Protestants, who in spite of predestination tried to ensure their salvation by working doggedly and methodically to achieve worldly success as a sign of divine favor. Furthermore, although Paracelsians like van Helmont railed against Aristotelian and Scholastic epistemology with its emphasis on reason, logic and disputation, they often used scholastic modes of reasoning and argumentation in their own work.[47]

Although deeply affected by the arguments of skeptics, Paracelsians and Helmontians never embraced skepticism. Instead of taking the positivistic, pragmatic approach to knowledge of constructive skeptics like Libavius, Mersenne, Gassendi, and Locke, they sought a new basis of absolute truth in intuitive knowledge, which, as committed esotericists, they continued to believe capable of revealing the essential nature and truth of things. These different epistemological approaches become a major issue in seventeenth-century debates about language and science. Paracelsus contended that the only way to discover the "signatures" or essential nature of things is through experience and experiment both within ourselves and in the external world. "He who wishes to explore nature," says Paracelsus in a favorite aphorism, "must tread her books with his feet."[48] This was hardly a novel idea. The concept of the book of nature as the word of God had been a commonplace of medieval preaching, intended for an illiterate audience. But Paracelsus gave the notion a radical, even revolutionary emphasis by dismissing all written authority. He encouraged individuals to interpret the book of nature on their own. Once again, one finds a strong parallel between this scientific approach and the Protestant emphasis on the individual's experience of God. Paracelsus believed that nature was constantly changing to meet the needs of new ages. Diseases and their remedies come and go and are found in one place and not another. Reliance on tradition is therefore useless and men must rely on their own investigations. Van Helmont shared this approach: "the Gods do fell Arts to Sweats, not to readings alone."[49] As historians have shown, however, the rejection of authority that is so much a part of the rhetoric of both Protestantism and the "new" science is basically just that, rhetoric. Both groups poured over ancient, medieval, and modern authorities in constructing their own "new" theologies and scientific theories.

The emphasis that Paracelsians and Helmontians placed on experimentation is another indication that their work was in line with the work of figures more commonly associated with the Scientific Revolution. Newman and Principe make this point in connection with George Starkey, whom they see as the major intermediary between van Helmont and Robert Boyle. They describe Starkey's methodical and systematic investigation of Helmontian chemistry, concluding that in terms of his laboratory practice he "seems to share more with the canonical figures of early modern science such as Galileo, Boyle, and Newton than he does with the popular image of the alchemist." They go as far as to claim that Starkey's notebooks could be taken as models of modern laboratory investigation.[50] The same cannot be said of van Helmont. For all the progressive elements in his work that laid the foundations for modern chemistry, there is a gap between his advocacy of experiment and his practice of it.[51] For example, criticizing those who objected to his book on the magnetic cure of wounds, he advised them to experiment before they dared to dismiss his theory:

> It is not agreeable to the custom of Naturalists, to argue from bare Authorities . . . make Tryall therefore, and call any of the recited examples [of magnetic and sympathetic effects] to the touch stone of experiment, that so you prove us guilty of falsehood; if you cannot, then at least come over to our side and believe them.[52]

There is not a modern scientist who would disagree with this injunction, at least until he or she understood the kind of crucial experiment van Helmont had in mind. A witch, he says, who has killed a horse by magic can be detected by burning the horse's heart impaled upon an arrow because her spirit will suffer the same intolerable sense of burning. He claimed that "the effect holds constantly good, and never fails to succeed upon experiment."[53] Van Helmont's certitude here may perplex or even amuse the modern reader, but it is important to realize how central the controversy about witches and witchcraft and the whole issue of what constituted valid evidence was in the early modern period. As we have seen, the debates over witches and what sort of evidence could be accepted as valid were crucial for determining the boundaries between the natural and the supernatural. The possibility and existence of witchcraft as well as other reported phenomena like second sight, the sympathetic cure of wounds, and the existence of fairies and elves raised profound questions about the authority and credibility of the Christian revelation, the role of God and spirits in the physical universe, and the epistemological problem of what constitutes sound scientific knowledge.[54] All these issues were deeply pondered by van Helmont, as they would later be by George Starkey, Robert Boyle and Isaac Newton. That these later scientists or natural philosophers came to different conclusions does not support the idea of a radical break between alchemy and chemistry but a more gradual theory of development.

As a reformer with an unshakable faith in his cause, van Helmont was neither modest nor moderate in setting forth his theories. Like most zealous innovators, he both anticipated his own martyrdom and helped to realize it. His personal sacrifice to the cause of science was not as great as some, but his philosophy did become a cross he was forced to bear. In a poignant passage, and one of the few in which he mentions his private affairs, van Helmont describes the death of his two eldest sons from the plague. He lays this disaster at the feet of his enemies, and accepts it with a fortitude bred from the unshakable conviction that he would be proven right in the end.[55]

Walter Pagel was one of the first of van Helmont's biographers to emphasize the profound importance his religious beliefs had on his scientific achievement.[56] The conviction that "ratio" was tied up with pride and atheism spurred his rebellion against Scholasticism and Galenism and turned him toward experiment and observation. This in turn led him to deprecate the theory of the four elements and the humors and to develop a dynamic theory of matter that envisioned diseases as specific entities attacking specific organs. His belief in God as the supreme and all-powerful creator caused him to scoff at materialistic explanations of natural events and to take account of psychical and psychological factors. His deprecation of reason led him to appreciate subconscious and autonomic nervous systems. Finally, his emphasis on specificity made him hostile to explanations based on astrological forces and analogies that were imprecise and unverifiable.[57] In these ways van Helmont laid much of the foundation for the chemical revolution of the eighteenth century. Van Helmont's religious bias was not wholly beneficial, however. His assumptions often colored his conclusions and prevented him from appreciating profoundly important discoveries such as Harvey's theory of the circulation of the blood or Copernicus's reorganization of the solar system. Pagel suggests that he was incapable of appreciating Harvey's discovery because of his inherent lack of interest in the application of mechanical laws to biological phenomena.[58] The same distaste for mechanistic or mathematical theories may have caused him to ignore Copernicus.

J. B. van Helmont's work reflects the contradictions and confusions of the unsettled time in which he lived. The Renaissance concept of man as a magus able to control and understand the universe retreated before Luther's and Calvin's pessimism and their insistence on man's sinful nature and abject helplessness. Van Helmont's work benefited from this conflict in attitudes. On the one hand man was still a potential magus who could look to the heavens and master the world around him. On the other, original sin made it more appropriate for him to fix his eyes firmly on the ground beneath his feet.[59] Van Helmont belongs in the Neoplatonic-Hermetic-Kabbalistic tradition that arose in the Renaissance, took refuge in various princely courts and academies during the turbulent period of the Reformation and Counter-Reformation, reemerging with renewed vigor in the seventeenth century.[60] His allegiance was to the esoteric tradition in Western culture, characterized by the following basic assumptions: (1) the universe is a organic whole with each part mirroring and corresponding to the others. (2) Every part of nature is alive and connected by a network of sympathies and antipathies. There is therefore essentially no difference between matter and spirit; they are entities along a single continuum. (3) The universe is not static but in a constant state of change and evolution, during which everything experiences transmutation as it is regenerated and eventually restored to its prelapsarian condition. (4) Human beings are made in God's image and share God's nature. Hence they play a pivotal role in restoring the world to its original perfection. (5) Thought and imagination are powerful forces that allow men to have access to different levels of reality and, through gnosis, to hidden, divine knowledge. (6) Inasmuch as both philosophy and theology arise from one source, God, they cannot be separated. They have existed from all time in a *prisca theologia*. Although partially obscured, this can be rediscovered so as to provide the basis for a truly ecumenical, peaceful, and harmonious world.[61] According to historians writing before 1960 it was precisely this worldview that was obliterated as a result of the onslaughts of the mechanical philosophy. What has become apparent since then, however, is that this worldview persisted, but it did not preclude many ideas and approaches that have come to characterize modern science such as quantification and the idea that theory must be subject to experimental testing and the possibility of falsification.

The acrimonious debate between certain Galenists and Paracelsian-Helmontians, while not to be underrated, belies the fact that over time the hostility diminished to the point that one author described himself as a "Chymical Galenist." A similar situation occurred in relation to the different schools of natural philosophy. While it is undoubtedly true that Aristotelian philosophy came under increasing attack in the seventeenth century, the idea that it was decisively routed from the philosophical and scientific scene is mistaken. Furthermore, while it used to be the accepted wisdom that Aristotelians distinguished themselves from atomists and mechanical philosophers, who, in turn, drew the line when it came to esotericists and vitalists, recent scholarship in history and the history of science has shown that the division between Aristotelians, mechanical philosophers, vitalists, and esotericists was not as clear-cut as once imagined; nor was there a straightforward correlation between specific scientific theories and corresponding religious and political beliefs.[62] Keith Hutchison's article, "What Happened to Occult Qualities in the Scientific Revolution," may appear to qualify this statement, for he argues that the recognition that occult qualities existed and were amenable to study separated occult and mechanical philosophers from Aristotelians and marked a significant advance in science. By identifying the occult with the insensible and hence incorporeal and spiritual, Aristotelians limited science to the study of what was apparent to the human senses, thus ignoring the enormous realm of insensible entities revealed by the microscope and telescope. The traditionally held belief that the Scientific Revolution entailed a turn toward the realm of sense experience and empirical observation therefore needs to be qualified. While experimental investigations were pursued, advances in science also came from what might be better described as a foray into "non-sense," a state of affairs that explains why the seventeenth-century naturalist Henry Power castigated Aristotelians for being "Sons of Sens" instead of experimental philosophers.[63]

In his recognition that occult forces can be studied even if they are imperfectly understood, van Helmont, a vitalist and occultist, as we have seen, was in what we with the advantage of hindsight might call the "progressive" camp. This, of course, was the argument Newton later used to defend himself against Leibniz's charge that gravity was an "occult" force and had no legitimate place in a scientific explanation. While progressive in this regard, van Helmont was also deeply influenced by the Aristotelian notion that form was an inherent aspect of matter. Hence it was clearly possible to combine aspects of Aristotelianism with esotericism. Evidence continually mounts to show that early modern natural philosophers had many different allegiances. Only this can explain the friendships and connections van Helmont had with scientists and philosophers with widely differing, yet often overlapping, views, such as Mersenne and Gassendi. It also explains the later interest shown in his work by Hartlib, Oldenburg, Comenius, Locke, Boyle, Leibniz, and Newton. Religion and science were thoroughly intermixed in the thinking of all these men, and whether they were vitalists, esotericists, or mechanical philosophers they shared the conviction that active, spiritual forces of some sort existed in matter.[64]

The epistemological dispute did not therefore lie primarily in admitting or rejecting spiritual or immaterial forces in nature but in agreeing as to exactly what kind of forces these were and how they worked. In this regard, it is important to emphasize that Boyle and many of his colleagues in England's Royal Society were deeply concerned by what they saw as the growth of skepticism and atheism and the role that natural philosophy may have inadvertently played in promoting both.[65] The debate about witchcraft was a case in point. During the seventeenth century a growing number of naturalistic and medical explanations had been given for the supposed actions of witches and spirits, and some natural philosophers like Boyle viewed these as a direct assault on Christianity. Hobbes, Descartes, and Spinoza were singled out as especially pernicious because by denying the existence of spirits, they were accused of undermining the belief in God. Joseph Glanvill, a vociferous advocate of the Royal Society, considered a disbelief in spirits the first step in the inevitable march to atheism. Like Boyle, Glanvill and the overwhelming majority of members in the Royal Society was deeply religious and took it for granted that the world was full of supernatural events and otherworldly forces. Not only did these men attempt to document their existence, but they wanted to explain their causes and significance. Boyle, as we have seen, subsidized the publication of the translation of an account of a poltergeist in France, and he corresponded with Joseph Glanvill about the desirability of collecting empirically verifiable accounts of witches and spirits. If the documents in Michael Hunter's *The Occult Laboratory* were indeed collected at the instigation of Boyle, we have further proof of his interest in supernatural and psychic powers, which we also saw in his involvement with the "stroker" Valentine Greatrakes. Boyle's endeavor to prove that spiritual and immaterial forces exist continued throughout his life and played an important part in his work as a chemist.

The mechanical philosophy did not triumph in the seventeenth century, nor did it triumph in the eighteenth. The extent to which the theories characteristic of Paracelsus and van Helmont continued to find adherents in the eighteenth century has become increasingly clear. In his pioneering work on Mersmerism, Robert Darnton shows that vitalistic theories like Mesmer's were eagerly embraced in France at the end of the eighteenth century, and the predilection for spiritualist cosmologies on the part of elite and popular culture alike prepared the way for Romanticism. Mesmer's critics recognized that his ideas were derived from Paracelsus, van Helmont, and Robert Fludd, but this awareness did nothing to decrease Mesmer's popularity since his ideas had a great deal in common with the theories about electricity, magnetism, gravity, light, and fire proposed by "respectable" authors. Mechanical explanations failed to give an adequate or even plausible account of the wondrous effects of these invisible forces, which were reported assiduously in the popular press. Chemists claimed to resurrect the forms of plants and animals by heating their calcinated ashes in hermetically sealed flasks, and books were published with ghostly images of these reconstituted beings.[66] Electrical charges were said to make plants grow and cure gout. A young boy supposedly regained the full use of his limbs after being thrown daily into a tub with a large electric eel—a cure we might well attribute to "natural" causes.[67] The fact that scientific and pseudoscientific explanations of the same phenomena were often so similar made it all the more difficult for ordinary people to distinguish fact from fiction. Goethe's Faust was a product of the eighteenth century, not the Middle Ages, and he appealed to an eighteenth century audience. People have always been drawn to the marvelous, and, paradoxically, the more science advanced, the more gullible and credulous a portion (at least) of the general public became. In his last book Carl Sagan lamented the fact that people preferred the pseudomarvels of pseudoscience to actual wonders of real science.[68] But as Robin Dunbar argues, real science is too difficult for most human beings, who have been programmed to have a "social" rather than a "scientific" brain.[69]

The difficulty most people have distinguishing between true and false science is hardly new; as we have seen, it was a central concern in the early modern period. The seventeenth century world traveler and diplomat John Finch wrote to his sister Anne Conway, deploring the fact that "the most part of Mankind being fooles and not able to judge, the knowledge of a few wise men will easily be outvoted by a Number of Talkers."[70] Finch suggests criteria for distinguishing truth from falsehood. He advises Anne to give credence only to those people who relate things on the basis of their own experience, who are known to possess good judgment, and whose opinion is not swayed by self-interest. But that, of course, was exactly the rub faced by people living at the time (as well as ever after). What counted for reliably reported experience and good, unbiased judgment, especially when what was at issue was invisible to the naked eye or senses? This was the problem Robert Boyle faced when he took up chemistry and began to study the work of his predecessors, particularly that of van Helmont largely as a result of his friendship with George Starkey, who based his own work on van Helmont.[71]

As a young man Boyle had been involved with members of the circle that developed around Samuel Hartlib, many of whom had adopted Helmontian theories. The idea that Boyle renounced Helmontian chemistry in favor of the mechanical philosophy when he moved to Oxford in 1655 has been effectively refuted by Antonio Clericuzio, along with Newman and Principle. It was only in the eighteenth century and largely as a result of Leibniz that Boyle was perceived as a supporter of the mechanical philosophy. This view of Boyle ignores the fact that, like van Helmont, he believed there were non-mechanical forces in the natural world such as spirits, seminal principles, and ferments, all of which had the power to fashion matter.[72] In his *Sceptical Chymist* Boyle makes frequent references to van Helmont through Carneades, his alter ego in the dialogue. So frequently does Carneades mention van Helmont that his companion Eleutherius remarks: "One would suspect . . . by what you have been discoursing, that you are not far from Helmont's opinion about the origination of composed bodies, and perhaps too dislike not the arguments which he employs to prove it."[73] John Ward, one of the members of Boyle's group in Oxford, wrote in his diary after reading the *Sceptical Chymist*: "Mr Boghil [Boyle] does mightily commend van Helmont."[74] Nevertheless in the *Sceptical Chymist* and certainly after its publication, Boyle did criticize some of van Helmont's ideas, for example, the notion that water was the source of everything.[75] But even at this point Clericuzio contends that "Boyle still regarded van Helmont's chemical ideas and experiments with the utmost interest," to such an extent, indeed, that he never rejected Helmontian chemistry as a whole.[76] Newman and Principe concur: ". . . Boyle's chymistry [*sic*] maintained Helmontian elements even in his late work."[77] What Boyle did was to reinterpret many of van Helmont's idea in terms of his corpuscular philosophy.[78] Boyle's contemporaries understood what he was attempting. In a letter to Oldenburg dated February 20, 1659/60, Robert Southwell writes: "I am extreame glad to understand that Mr Boyle is ingaged in soe advantagious a designe as ye collation of Chymicall experiments with Atomicall notions. He is a person of soe much ability, and soe much caution withall, that I doe not think he would have publisht unto you his designe without haveing made some successfull and happy essay therein allready."[79] Boyle publicly announced his intentions in the preface to the fourth edition of his *Certain Physiological Essays* (1661), where he proposed "the Desirableness of a Good Intelligence betwixt the Corpuscularian Philosophers and the Chymists."

The Restoration did not therefore bring about a crisis in the English Helmontian movement, as some scholars have asserted. Helmontian iatrochemistry became even more popular after 1660.[80] It enjoyed the strong support of Charles II, George Villiers, Duke of Buckingham, Prince Rupert of the Rhine, and the Archbishop of Canterbury, among others. The Helmontians were reinforced by the arrival in England of two German iatrochemists, Joachim Polemann and Albert Ortho Faber. Polemann's *Novum lumen medicum* (Amsterdam, 1659) was translated into English in 1662 and gave a clear exposition of Helmontian medicine. Faber became Charles II's personal physician in 1661. He was a Quaker, which supports Rattansi's contention that religious radicals were drawn to van Helmont. John Chandler, another Quaker, translated van Helmont's *Ortus medicinae*. In "The Translator's Premonition to the Candid Reader," he accepts van Helmont's claim that true physicians are made by God, not by the schools.[81]

The "annus mirabilis" for English Helmontians was 1665, for in that year Marchamont Nedham's *Medela medicinae* was published, which was essentially a manifesto for Helmontianism. In that same year the "Society for Chemical Physicians" was planned, and although it never materialized, it provides evidence of the significant influence van Helmont had on later generations of chemists and alchemists. Because of its harsh attack on Galenic medicine, a pamphlet war followed the publication of Nedham's book with such ripostes as Robert Sprackling's *Medela ignorantiae* (London, 1665) and John Twysden's *Medicina vetrum vindicate: Or, an Answer to a Book entitled Medela medincinae* (London, 1666). But there were those like George Castle, a Fellow of the Royal Society, who took a more conciliatory approach, as the title of his contribution to the debate makes clear: *The Chymical Galenist* (London, 1667). As he says:

> My design is, to shew that though the Physiology and Pathology of physick, ought to be modeled according to the new Discoveries in anatomy, and the Democriticial and Chymical Principles, yet that many of the Rules, Methods, and medicines, which more immediately respect the useful and practical part, are still to be retained, and that they are rather more reconcileable to the Moderns, than they were to the Ancient Hypotheses (sig. A3v-A4r).

William Johnson also promoted compromise claiming that "the Judicious and Learned do not build the Praises of Galen, on the disgrace of Vanhelmont, but honour both according to their respective worth."[82] By 1670s even such a fierce critic of Galen and Galenists as William Simpson had mitigated his polemical tone. This suggests that iatrochemistry had now established itself to the point that it didn't need to defend itself so aggressively. As Clericuzio says, by the 1670s "Helmontian theories, . . . interpreted in the light of the particulate theory of matter had become an integral part of natural philosophy."[83]

The integration of Helmontian theories into natural philosophy helps to explain the number of Helmontians who became members of the Royal

Society, such as William Simpson and John Webster. Simpson wrote a tract on fermentation, which he dedicated to Lord Brouncker, who was both the President of the Royal Society and patron of Helmontian chemists.[84] In his *Metallographia, or a History of Metals* published in 1671 Webster took the Helmontian position that metals grew in the earth from seeds. Dr. Daniel Foote was another Helmontian member of the Royal Society. Henry Stubbe may not have been off the mark when he complained to Boyle that the Royal Society was subverting traditional learning by taking the side of Helmontians.[85] One can only conclude that iatrochemistry did very well in Restoration England, so well, indeed, that it may have influenced the greatest scientist of the age, Isaac Newton.

Exactly how profound an influence van Helmont had on Newton is hard to say, for Newton was an avid reader of a wide variety of alchemical/chemical treatises. We know that Newton owned van Helmont's *Ortus Medicinae*, and according to Clericuzio he read it.[86] John Maynard Keynes famously called Newton the "last of the Magi" because of his intense interest in alchemy, an interest well documented through the scholarship of Betty Jo Dobbs, Richard Westfall, and others. Newton wrote some one million, five hundred thousand words on alchemy, far more that he ever expended on the physics for which he is remembered. Alchemy and biblical interpretation were Newton's primary occupations, a fact that perplexed earlier Newton scholars, who dismissed both as irrelevant to Newton's scientific genius. But as Dobbs, Westfall, and Goldish have shown, the years Newton spent studying the Book of Daniel and Revelations and pouring over alchemical books and manuscripts in the laboratory he set up in Cambridge were of the utmost importance in shaping his ultimate view of the mechanics of the universe and his concept of gravity.

With the publication of Pierre Nicolas Lenglet de Fresnoy's *Histoire de la philosophie hermétique* (1742), the position associated most closely today with Jung, namely that alchemy had nothing to do with real laboratory chemistry but involved the spiritual and psychological transformation of alchemists themselves, was forcefully presented. This interpretation was reinforced by nineteenth-century spiritualists and theosophists, and by Freud's student Herbert Silberer. Jung followed this tradition. While he offered insights into the spiritual and psychological aspects of some alchemical texts, his interpretation failed to appreciate the development of laboratory and experimental alchemy over the long term and the diversity of alchemical theories and practices in different periods. Furthermore, Jung mistakenly assumed that vitalism was an essential characteristic of alchemy, ignoring the existence of materialistic and corpuscular theories.[87] Jung's interpretation is further undermined by the fact that more recent scholars of alchemy have decoded alchemical texts to reveal descriptions of actual laboratory practices. However fanciful alchemical descriptions may appear to the uninitiated reader, scholars like Dobbs, Newman, and Principe have successfully cracked their codes, which is, of course, precisely what Newton managed to do on many occasions as he collected, read, and reread alchemical texts. Finally, the images alchemists' used to symbolize their work were not necessarily eruptions from some pre-existing unconscious, as Jung claimed, but metaphors and symbols consistent both with the kind of "emblematic world view" characteristic of much of alchemy and with actual observations of chemical reactions, a point made by both Dobbs and Principe. We can leave the last word, however, to Lavoisier himself. As he wrote in *Essays Physical and Chemical* (1776):

> We are astonished, in reading this Treatise [van Helmont's *Ortus*], to find an infinite number of facts, which we are accustomed to consider as more modern, and we cannot forebear to acknowledge, that Van Helmont has related, at that period, almost every thing, which we are now acquainted with, on this subject. . . . It is easy to see that almost all the discoveries of this kind, which we have usually attributed to Mr. Boyle, really belong to Van Helmont, and that the latter has even carried his theory much farther.[88]

This chapter demonstrates on a concrete and individual level the impossibility of separating distinct strands of religion, magic, and science in the work of three men recognized for their contributions to the development of modern science, Jan Baptista van Helmont, Robert Boyle, and Isaac Newton. What becomes apparent is that our categories and definitions of religion, magic, and science do not fit the way people viewed the world in the early modern West. Given the fraught history and contentious nature of the way these terms have been defined, they may not even fit the actual thinking of most people in the world today, but that is another subject.

EPILOGUE

Social scientists and intellectuals in the West have been forecasting the end of religion and magic and the triumph of science for some three centuries. In 1822 Thomas Jefferson echoed what had by then become a truism when he wrote, "I trust that there is not a young man now living in the United States who will not die a Unitarian," by which he meant die a member of a denomination that had removed every hint of the superstitious and the miraculous from religion.[1] The fact that Unitarianism actually declined in the nineteenth century while more traditional and conservative religions flourished did nothing to alter the perception among many intellectuals that secularization was the dominant characteristic of modernity, an unstoppable force that would ultimately prevail.[2] By the 1970s secularization theory was the reigning dogma. It was not until the late 1980s that the theory was criticized on the grounds that it was an ideology, rather than a statement of fact, based on bad history and the mistaken notion that the past was more pious than the present.[3] It is the purpose of this epilogue to reflect on previous chapters and describe some of the ways in which secularization theory proved spectacularly wrong and to make some suggestions as to why scholars and intellectuals were misled. For example, the idea that the emergence of science required the demise of religion and that science was consequently a major factor in secularization is quite simply wrong, as we have seen. The rise of science in the West went hand in hand with religion and cannot be understood without acknowledging the important contributions made by a wide spectrum of religions' beliefs from orthodox, unorthodox, and esoteric sources. Another piece of accepted wisdom is equally invalid, namely that the Middle Ages was the "Age of Faith" while the early modern and modern periods represented "the disenchantment of the world," to use Weber's phrase once again. Finally, the idea that rationality would inevitably replace "superstitious religious beliefs" once science was ascendant and secularization in place has not proven true. On the contrary, religious and "magical" beliefs have proliferated and fundamentalist religious doctrines are more powerful and powerfully organized than ever before.

The Enlightenment attack on religion laid the foundation for the view that religion was a human construct arising from fear and anxiety. This view was carried forward into the nineteenth-century evolutionary schemes of anthropologists like Edward B. Tylor and James Frazer, who believed that religion would be superseded by science, and in Freudian, Marxist, and sociological analyses of religion. Max Weber's theory of the religious roots of capitalism and the consequent "disenchantment of the world" appeared to be a further nail in religion's coffin, but it was science that was expected to provide the final coup de grâce. As secularization theory presumed, the influence of religion would decline as societies became increasingly organized on rational principles. Anthony Wallace proclaimed this message in 1966 with embarrassing certitude:

> . . . as a cultural trait, belief in supernatural powers is doomed to die out, all over the world, as a result of the increasing adequacy and diffusion of scientific knowledge and the realization by secular faiths that supernatural belief is not necessary to the effective use of ritual. The question of whether such a denoument will be good or bad for humanity is irrelevant to the prediction; the process is inevitable.[4]

Recent events and recent scholarship have proven Wallace and secularization theory in general categorically wrong on two counts: first, with the assertion that science and religion are incompatible and that the development of the first necessarily entails the elimination of the latter; and second, with the conviction that, given the chance, human beings would embrace scientific rationalism because it would provide for all their intellectual, material, emotional, and spiritual needs.

As we have seen, contemporary historians of science dismiss the once dominant idea that modern science and modern ways of thinking emerged when rationalist philosophers rejected irrational religious and magical thinking. We are now aware of the important way in which Christianity encouraged science. The notion of the divinely created universe as "a book of nature" that must be studied in conjunction with Scripture to reveal the majesty and goodness of God, the idea of history as linear and progressive, the role of man as nature's steward, and, most importantly, the idea that man was created in the image of God all contributed to legitimizing and promoting scientific activity. However, as scholars have increasingly realized, the term "Christianity" is incapable of conveying the multiple religious, occult, and magical beliefs of Europeans and Americans in the past as well as the present. This realization has led to a fundamental reevaluation of the nature of Western religious practice and belief by showing that religious identities were never fixed along simple denominational lines. The upshot of all this is that modern science and modern thought have a far more complex ancestry than was once believed. The view of a triumphal "scientism" at the expense of religious and "superstitious" thinking, however defined, does not fit the facts. Science developed in tandem with "religious" ideas taken in the broadest sense. The conflict between religion and science that used to be taken as axiomatic was not present for long periods of Western history. Conflict arose at certain times for certain identifiable reasons. For example, in the first Christian centuries when Christianity was struggling to establish itself, Christian theologians rejected classical thought. "What indeed has Athens to do with Jerusalem?" queried the second-century Church Father Tertullian.[5] Once recognized as the religion of the Roman Empire, however, Christianity had no problem borrowing from classical philosophy, as, indeed, it had to if it was to create a coherent philosophical and cultural world view. The trial of Galileo was another instance in which religion and science came into conflict, but this had more to do with Reformation and Counter-Reformation politics than any real antipathy for Copernicanism itself. (It must be remembered that Galileo's trial came almost a century after the publication of Copernicus's treatise.) Evolution created another potential conflict between religion and science, but primarily for fundamentalists. Mainstream religions, including Catholicism, eventually accepted evolution with the proviso that human beings are exempt, at least in regard to the soul.

It is not simply that modern science developed in the context of religious and magical thinking, but that religious and magical thinking have prevailed in all epochs of Western history, including the present. The idea of the Middle Ages as a period of fervent devotion severely distorts the historical picture because it suggests that "faith" was synonymous with Catholicism and that "Catholicism" could be defined as a determined set of doctrines and rituals, neither of which was the case. Doctrine and ritual were subject to ongoing revision and restatement. For example, the doctrine of the Trinity was not formulated until the Council of Chaldea in the sixth century CE, and even then the Arian wing of the Church refused to accept it. The doctrine of transubstantiation, which was to become a key component of Catholic doctrine and faith, was not promulgated until the Fourth Lateran Council in 1215, and scholars have argued that the doctrine created severe skepticism on the part of many Catholics, an example of another fault line in the "Age of Faith" moniker.[6]

Attempts to reconstruct the actual beliefs of medieval and early modern Europeans reveal that magic and superstition predominated. Even priests, especially those on the parish level, were woefully ignorant of Catholic theology and accepted the folklore and magical beliefs of their parishioners. A very good indication of just how widespread magic continued to be in the early modern period is contained in the visitation reports of Protestant and Catholic reformers, who made a concerted effort to question people about their religious beliefs. This is not to claim that the same kind of magical beliefs found in the early modern West have continued to exist in the more economically and scientifically advanced parts of the contemporary world, although similar beliefs and practices clearly do in less-developed countries. But even in the developed world, religious and magical thinking has not disappeared with the emergence of modern science, as secularization theory predicted it gradually would. How can we account for this? In *The Trouble with Science* Robin Dunbar maintains that human beings are far more fascinated with the social than the physical world, that, in fact, the vast majority of people view the world anthropomorphically. We routinely talk about such things as a "vicious" wind and a "raging" sea. We see faces everywhere, not just in the moon. As a result of the "pathetic fallacy" nature appears to respond to our moods: the sky darkened when Jesus died even though it was daylight. In Romantic poetry landscapes mirror their authors' emotions. We are primarily social animals with social brains and are far more interested in group dynamics and gossip than scientific abstractions. Science, according to Dunbar, is "unnatural" inasmuch as it substitutes logical rigor, mathematic models, and probabilities for common sense and the senses in general. Human beings want certainties and absolutes, not probabilities; and religions, particularly fundamentalist religions, provide these, which is a great part of their attraction.[7] But even more than the sheer difficulty of science, what accounts for the persistence of religion is that it, and only it, can provide what most human beings seem to want so desperately: explanations for why bad things happen to good people and the promise of some kind of ultimate reward and compensation for the inevitable pain and suffering of human life. When secularization theory was still regnant, Melford Spiro claimed that religion would decline proportionately as science, technology, and culture increasingly fulfilled human intellectual, material, and cultural needs:

> I have stressed, with respect to the three sets of [human] desires [cognitive, substantial, expressive] . . . that in the absence of alternative institutional means, it is religion which is the means par excellence for their satisfaction. If cognitive desires, for example, are satisfied by science; if substantive desires are satisfied by technology; or if expressive desires are satisfied by politics or art or magic, religion should, by that extent, be less important for their satisfaction. In short, the importance of religion would be expected to vary inversely with the importance of other, projective and realistic institutions.[8]

The problem is that neither science nor culture can provide the most important things that religion promises. This is the premise of Rodney Stark and William Bainbridge, who claim that "Humans have a persistent desire for rewards only the gods can grant, unless human become gods."[9] In their view secularization is a process that must necessarily lead to religious renewal because a secular world cannot answer life's great existential questions or provide mortal, fallible human beings with a sense of ultimate security. By proposing a series of axioms and deriving propositions from these, they make a compelling case that religion is not going away in the foreseeable future because: (1) human beings exist in time and want to influence the future; and (2) the rewards they seek go beyond things that can be obtained or even known by science or acquired through the accumulation of material goods. "However healthy science makes us, we cannot live forever. However wealthy technology makes us, relative deprivation will always exist."[10] People are willing to pay great costs for great rewards, and if the rewards they want do not exist in this world, they are willing to accept explanations claiming they will materialize in a future time and place. Hence their willingness to embrace religion.

In light of contemporary events, rethinking secularization is a timely occupation. The decline of religion and the embracing of rationality has not occurred to the extent once predicted. Scholars who were themselves products of enlightenment thinking were blinded by their own preconceptions into concluding that these could and would become universal. They also mistook the decline in mainstream religions for a decline in religion itself, failing to realize how creative human beings are when it comes to refashioning religion to suit changing times and needs. Roxanne Euben contrasts the decline in voter turnout in secular democracies to the rise of religio-political movements around the world bent on reforming the world:

> . . . why is it that secular liberal democracies such as the United States are witnessing sharply declining rates of voter turnout and increasing alienation from politics at the very moment that religio-political movements are galvanizing people into extraordinary attempts to remake the political world?[11]

Low voter turnout is a problem in liberal democracies, but the injection of religion into politics that has become so noticeable in the United States in recent years has galvanized religious conservatives in the attempt to impose their convictions on the country as a whole. The Ten Commandments do not offer the complex framework of Sharia law, but many Christian conservatives appear to take them as a basis for some kind of universally applicable religiously based law. While an individual's religious views will inevitably influence his or her politics, few people are fully aware of how dangerous the mix of religion and politics can be when absolute convictions make any kind of compromise an impossibility.[12] If we look back to pre-Hitler Germany, one can find the same mix of politics and religion. Fritz Stern attributes Hitler's success to this explosive mix: "Some people recognized the moral perils of mixing religion and politics but many more were seduced by it. It was the pseudo-religious transfiguration of politics that largely ensured his [Hitler's] success, notably in Protestant areas."[13] Stern credits the longing for a sense of community predicated on a kind of religious authoritarianism for the triumph of fascism. He singles out the role that anti-modernism played in bringing Hitler to power. Modern religious conservatives are dangerously close to advocating the same kind of communal religious authoritarianism. The conservative columnist Cal Thomas, for example, excoriates the Enlightenment and compares it unfavorably to the Reformation:

> The Enlightenment was anything but enlightening. It was a reaction by European intellectuals to the Reformation period that preceded it. Unlike the Reformation, the Enlightenment thinkers believed that people and society were perfectable. They rejected the existence of (or accountability to) a creator God. . . . As an autonomous creature, Man was then free to establish his own laws and morals and to change them at any time to suit the prevailing philosophical and pragmatic winds of the age.

Thomas wants to expel "the bogus philosophy of the Enlightenment" and re-establish "the Reformation philosophy it replaced."[14] He is apparently unaware of how deadly and divisive the religious conflicts marking the Reformation were. Only during the sixteenth and seventeenth century did both churches and states attempt to control the minds and thoughts of their subjects, not simply their actions. The early modern period was the age of catechisms, the Consistory at Geneva, the Inquisition, Auto da fés, the Index, religious wars, civil wars, witch hunts, and witch burnings. It is understandable that by the end of the seventeenth century, many people decided that religion was the problem, not the solution, as Cal Thomas would have it.

Thomas follows in the footsteps of earlier Enlightenment critics, who castigated the *philosophes* for replacing God with Reason and thus starting humans on the inevitable course to the ovens at Auschwitz. As Horkheimer and Adorno wrote in 1947 with the Holocaust clearly in mind, "The Enlightenment had always aimed at liberating men from fear and establishing their sovereignty. Yet the fully enlightened earth radiates disaster triumphant."[15] In their reading of the past there was no reason for the West to be proud of its success in science and technology, for "[o]n the road to modern science, men renounce any claim to meaning. They substitute formula for concept, rule and probability for cause and motive."[16] The instrumentalization of reason—by which they meant the use of reason without any moral or ethical constraints or parameters but simply to solve any given problem—was responsible for the rise of totalitarianism, antisemitism, racism, intolerance, and ultimately the extermination of human beings.

The attack on the Enlightenment came, and still comes, from the left as well as the right, from postmodernists who view the claim for reason, tolerance, and universal brotherhood as a cynical ploy to mask Western domination and exploitation. Postmodernists jumped on the anti-Enlightenment bandwagon with gusto, accusing it of laying the foundations for rationalism, instrumentalism, scientism, logocentrism, universalism, masculinism, and everything that could be summed up negatively as "moderism."[17] According to their analysis by denigrating history and tradition, the Enlightenment placed an undue emphasis on human reason to reform society; it established universal rules and constrained individuals by force if necessary to conform to abstract principles of goodness, justice, and truth devised by humans without any external source of validation; and, finally, the enlightenment view of rationality turned technology, science, and knowledge in general into commodities disconnected from wisdom and ethics. Thus instead of bringing the sweet dreams of reason, the Enlightenment brought war, famine, disease, persecution, enslavement, the subordination of women, and ecological disaster.[18] One critic went so far as to claim Hitler as a purveyor of Enlightenment values.[19] What is so interesting about this critique is, first, that it is based on an outmoded view of the Enlightenment as a homogeneous phenomenon, and, second, and more importantly, it judges enlightenment thinkers by the standards they themselves devised. Both points should give us pause.

It may not be fair to criticize earlier critics of the Enlightenment like Horkeimer and Adorno because they wrote before the surge of new Enlightenment scholarship that has appeared in the past few decades. Their critique was based on the view of the Enlightenment established primarily by Ernst Cassirer but by other intellectual historians as well in the 1930s, 1940s, and 1950s. Casssirer characterized the Enlightenment as a period dominated by "a few great fundamental ideas expressed with strict consistence and in exact arrangement," to quote the introduction to his famous work, *The Philosophy of the Enlightenment* (1932).[20] Cassirer emphasized the "universal" nature of the Enlightenment and the place of science in demonstrating universal laws, which carried over into the ideal of a universal legal code, ideal forms of government, and even an ideal of happiness. For Cassirer, writing as a liberal in defense of Weimar values under the threat of National Socialism, reason and science were man's most valuable assets, and the age which championed both was worthy of respect and emulation: "The age which venerated reason and science as man's highest faculty cannot and must not be lost even for us. We must find a way not only to see that age in its own shape but to release again those original forces which brought forth and molded this shape."[21]

Writing after the Holocaust, Horkheimer and Adorno saw the use to which reason and science had been put and quite naturally, though quite wrongly, as defenders of the Enlightenment argue, blamed the century and the intellectuals responsible for placing so much faith in reason and science. As more modern studies of the eighteenth century have shown, however, the period was far more diverse and varied than once imagined. What modern scholars see as the primary characteristic of the Enlightenment is the fragmentization of opinion in every conceivable area. Historians now view it as a period in which new forms of media combined with new institutions (academies, universities, reading societies, salons, journals, newspapers, translations, lending libraries) to create a cacophony of competing and conflicting information supporting a proliferation of religious and political ideologies from the most conservative to the most radical.[22] For example, the attack on religion and emphasis on reason was only one aspect of enlightenment thought championed by specific individuals for specific reasons in specific places, but not by the public at large. In fact, religion has returned to the Enlightenment with a vengeance, or one might say more accurately that religion never left except in the imagination of those who feared it had or who had a polemical reason for claiming it had. To quote the *Edinburgh Magazine* (1758): "There never perhaps was an age in which religion was so much in fashion among us, as it has long been . . . [G]reat is the thirst of multitudes after little refined points and particular doctrines of piety."[23] While it is true that there was an attack on religion, especially in France and largely as a result of the close alliance between the French Church and the State, there was no such attack in England or in most other European countries, not to mention America. Even in France, religion did not disappear. It was, however, transformed as a new kind of inner and subjective religiosity (reflected in Pietism, Methodism, and other forms of esoteric belief like Rosicrucianism, Freemasonry, Swedenborgianism, and Mesmerism) took the place of formal ritual observance.[24] The eighteenth century was the period in which piety was privatized and firmly centered in beliefs rather than practices, although it continued to be a defining factor in social and political life.[25] Those who criticize the Enlightenment for rejecting metaphysics for mathematics and divorcing reason from ethics overlook the fact that Enlightenment thinkers were much closer to the wars of religion than we are and consequently had reason to be wary of the terrible consequences of religious conflict. The idea of a formalized analytical mathematics or calculus that could be applied to problems logically without philosophical or religious qualifications undoubtedly seemed like a worthy goal to these academicians. Mary Terrall quotes Condorcet's "Reception Speech," when he was inducted into the Académie Française in 1782 (after a bitterly contested election). It is difficult to see anything problematic with the universalism he proclaimed in terms of the capacity that every person has to become enlightened: "The method of discovering truth has been reduced to an art, one could almost say to a set of formulae. Reason has finally recognized the route that it must follow and seized the thread that will prevent it from going astray. . . ."[26]

Postmodernists pride themselves on taking a historicist position and understanding how history influences what people think and do. Enlightenment thinkers were already aware of this fact, however, and the kind of universalism many of them advocated was offered to counter the centripetal forces encouraged by the growth of individualism, consumerism, and a persisting localism, which they recognized and in many cases deplored. Enlightenment thinkers were not uniformly self-satisfied, smug, imperious, misogynist, and racist, although some of them were. As early as the sixteenth century in his essay *Of Cannibals*, Montaigne questioned what it was to be European, suggesting that Europeans were not superior to cannibals but in many ways their inferiors, especially when it came to religion and the treatment of their fellow human beings. Defoe makes it clear that Friday was bewildered by ideas and behavior that seemed utterly normal to Robinson Crusoe. The Houyhnhms found Gulliver's descriptions of European religion and politics incomprehensible. Terry Castle goes so far as to claim that "eighteenth-century culture as a whole might . . . be termed, without exaggeration, a culture of travesty" and that eighteenth-century English society was "a world of masqueraders and artificers, self alienation and phantasmogoria," a view seconded by many living at the time.[27] One very important effect of the various disguises individuals assumed in the new "enterprise society" was to demonstrate that nature was not natural but a product of custom and convention. What many people considered "natural" in terms of gender and class, for example, was shown to be artificial and socially constructed.[28] Peter Reill discusses the historicism of German enlightenment thinkers, attributing it in part to Leibniz's theory of the monad as a dynamic entity constantly changing and developing and wholly unlike the lifeless atoms of the mechanic philosophers.[29]

A final way in which the Enlightenment has been misunderstood is reflected in Foucault's pessimistic view that it was a period in which individuals were forced to adapt to prescribed codes of behavior and imprisoned or institutionalized as criminals, madmen, and deviants if they did not. While there are elements of truth in Foucault's analysis, it ignores the humanitarian values that motivated so much of enlightenment thought. What lies at the root of Foucault's criticisms like all the others is the admittedly lamentable fact that many enlightenment thinkers as well as those who followed in their footsteps did not always live up to the promises implied by the notions of equality, liberty, and fraternity, that too often human self-interest got in the way and what started off well ended badly. However, there is no escaping the fact that it takes enlightenment values to criticize the Enlightenment, and this must never be forgotten. The Enlightenment provided human beings with the tools to be self-reflective and self-critical. As Jürgen Haberman insists, "it is the very nature of the Enlightenment to enlighten itself about itself, and about the harm that it does."[30] Although a student of Adorno and Horkheimer, Habermas has a much more positive view of the Enlightenment, seeing it as a period in which the public sphere developed as a place in which critical reason and reflection could function and develop.[31] Habermas emphasizes the fact that the Enlightenment remains unfinished. The reach of Enlightenment thinkers may have exceeded their grasp, but their ideals are still worthy of implementation.

What is so discouraging for anyone who agrees with Habermas and embraces the ideal of pluralism, tolerance, and the freedom of inquiry that marked enlightenment thinking at its best is that many of those who reject these values in the name of some absolute, revealed religious truth have no comprehension of the dangers implicit in their attitude and actions, not only for those they dislike but for themselves as well. On that subject, I think the last word can be left to Azar Nafisi. As she makes clear in her book, *Reading Lolita in Teheran*, however misapplied or misinterpreted enlightenment ideals may be, they are the only ones possible in an increasingly global and pluralistic world. Only by taking enlightenment ideals seriously are we able to judge a situation in which they do not exist. It will have to be seen if these ideals, forged out of the debates about religion, magic, and science that took place in the early modern period, can prevail or even survive in our current polemical religious and political climate.

Notes

INTRODUCTION

1. King, *Introduction to Religion*, 64.
2. Pettersson, "Magic—Religion."
3. Latour, *The Pasteurization of France*, 216.
4. For the tendentious ways the terms religion, magic, and science have been used, see Styers, *Making Magic*.
5. Geertz, "Towards an Interpretive Theory of Culture," 5.
6. Swift, part 3 of *Gulliver's Travels*.
7. Struever, *The Language of History in the Renaissance*.
8. Cited in Ricken, *Linguistics, Anthropology and Philosophy*, 221. Burke and Porter, ed., *The Social History of Language*; idem; *Language, Self, and Society*.
9. McManners, *Death and the Enlightenment*, 6.
10. Cox, *The Parish Registers of England*, 175.
11. *Diary of Samuel Pepys*, February 21, 22, 25, 1665.
12. *The Life and Times of Anthony Wood*, II, 68.
13. Cockaye, *Hubbub*.
14. Clarkson, *Death, Disease & Famine*, 39.
15. Ibid., 96.
16. *Thesaurus Novus Experimentiae Medicae Aurens: oder Guldener Artzney-Schatz neuer niemals entdeckten Medicamenten wider allerhand Leibs-Kranckheiten; auss den fürtrefflichen Schriften Fr. Merc. Van Helmont, Rob. Boyle, und anderer beruhmten Medicorum....* (Basel, 1723).
17. Peter, "Disease and the Sick at the End of the Eighteenth Century," 114.
18. Thomas, *Religion and the Decline of Magic*, 10.
19. Ibid., 177.
20. Aubrey, *Brief Lives*, 14–15: " . . . St John's night, 1673, in danger of being run through with a sword by a young gallant at Mr Burges' chamber in the Middle Temple. . . .[T]he year that I lay at Mrs. Neve's; for that time I was in great danger of being killed by a drunkard in the street opposite Gray's Inn gate—a gentleman whom I never saw before, but (thanks be to God) one of his companions hindered his thrust."

21. Webster, *Health, Medicine and Mortality in the Sixteenth Century.*

22. Underdown, *Fire from Heaven*; Roberts, "Agencies Human and Divine."

23. Polyani, *The Great Transformation.*

24. Ricoeur, *The Symbolism of Evil*, 281.

25. Dagens suggested this rubric applied to Catholic France, but it is even truer for Protestants. See *Bérulle et les origins de la restauration catholique.*

26. Cited in Harrison, *The Fall of Man and the Foundations of Science*, 31.

27. Ibid., 3.

28. Ibid., 182. Cf. Harrison, *The Bible, Protestantism and the Rise of Natural Science.*

29. Sarton, *Introduction to the History of Science*, 1: 19.

30. Gay, *The Enlightenment*. Vol. 1: *The Rise of Modern Paganism.*

31. Sheehan, "The Enlightenment, Religion, and the Enigma of Secularization."

32. Cf. Saler, "Modernity and Enchantment."

33. Koyré, *From the Closed World to the Infinite Universe.*

34. It is worth quoting Keynes assessment of Newton at some length because it is so different from the usual view of the man. "Newton was not the first of the age of reason. He was the last of the magicians, the last of the Babylonians and Sumerians, the last great mind which looked out on the visible and intellectual world with the same eyes as those who began to build our intellectual inheritance rather less than 10,000 years ago. Isaac Newton, a posthumous child born with no father on Christmas Day, 1642, was the last wonderchild to whom the Magi could do sincere and appropriate homage."

35. Cohen, *The Scientific Revolution*, 499.

36. See Céard, *La Nature et les prodiges*; Greenblatt, *Marvelous Possessions*; Eamon, *Science and the Secrets of Nature*, 108; Thorndike, "Newness and Novelty in Seventeenth-Century Science and Medicine."

37. Ashworth, "Natural History and the Emblematic World View," 318. Not all scholars agree that the new world had such a profound effect, contending that the only impact was to confirm old world prejudices. Both views are correct, but they apply to different groups and individuals. Cf. Elliott, *The Old World and the New*; Gliozzi, *Adamo e il nuovo mondo*; Ryan, "Assimilating New Worlds in the Sixteenth and Seventeenth Centuries"; Grafton (with Shelford and Siraisi), *New Worlds, Ancient Texts.*

38. Slaughter, *Universal Language and Scientific Taxonomy in the Seventeenth Century*, 57.

39. Popkin, *The History of Scepticism.*

40. Russell and Edwards, *Why I am not a Christian*, 107.

41. Bouwsma, *A Usable Past*, 2.

42. Trevor-Roper thought the witch hunts posed an "intellectual challenge" inasmuch as they occurred during the period said to usher in modern rationality (*The European Witch-Craze*); Tambiah was also puzzled (*Magic, Science, and Religion*, 47).

43. As scholars have shown, Bacon was not the naïve inductionist he presents himself as being, nor was Newton adverse to hypothesizing. Many of the "facts"

reported in early modern scholarship are also highly questionable, as the reader will discover.

44. This has nothing to do with personal piety, which probably did increase in the period and was perfectly consistent with secularization. Hunter, *Science and Society*, ch. 7; idem, "Science and Heterodoxy"; Bouwsma, "The Secularization of Society in the Seventeenth Century."

45. The interest in world history in recent years has shown that there is no such thing as "modernity" in the singular, only multiple modernities. The same can be said for early modernities. This book deals with developments in Western Europe and American from 1450–1800 and has no pretensions to make universal claims. Cf. *Daedalus* "Early Modernitites" 127 (Summer 1998).

CHAPTER 1

1. Yates, *The Art of Memory*, 132.
2. Ibid., 133.
3. Wittkower and Wittkower, *Born Under Saturn.*
4. Ashworth "Natural History and the Emblematic World View"; Gombrich, "Icones Symbolicae."
5. Foucault, *The Order of Things*, 27.
6. Ibid., 17.
7. Foucault, *The Order of Things*, 32.
8. Pico della Mirandola, "Oration" (http://www.fordham.edu/halsall/med/oration.html).
9. Brown, *Augustine of Hippo*.
10. According to Aristotle, every created thing has a specific end or purpose (*telos*), what we would describe today as its own DNA. An acorn's *telos* is to become an oak, just as a child becomes an adult.
11. Smith, "The Influence of Symbols on Social Change."
12. Ibid., 143.
13. Recent scholars question whether Luther actually said, "Here I stand. I can do no other," since these words were only inserted in later versions of the speech. But the fact that they were added at all supports the point I am trying to make.
14. Cited in Harrison, *The Bible, Protestantism, and the Rise of Natural Science*, 179.
15. The consequences of Copernicanism took time to sink in. Westman concludes that before 1600 there were only ten European thinkers who accepted the physical truth of Copernicus's theory. But if the acceptance of Copernicanism was delayed, there were many other cracks developing in the Ptolemaic-Aristotelian worldview that undermined traditional assumptions: for example, the observations and theories of Tycho Brahe and Johannes Kepler; Galileo's discovery of spots on the sun, craters on the moon, and the moons of Jupiter, all of which undermined the idea of the heavens as perfect and unchanging, plus new ideas about meteors and comets.

16. *Pensées*, 434.

17. *Pensées*, 72.

18. *Pensées*, 10.

19. *Pensées*, 347.

20. Bouwsma, "Anxiety and the Formation of Early Modern Culture," 157–189.

21. Allen, "The Degeneration of Man"; Williamson, "Mutability, Decay, and Seventeenth Century Melancholy"; Harris, *All Coherence Gone.*

22. Burnet, *Theory of the Earth*, Epistle Dedicatory.

23. Cited in Hayden, *The Counter-Renaissance*, 13.

24. Greenblatt, *Marvelous Possessions*, 14, 20.

25. Ibid., 21.

26. Slaughter, *Universal Language*, 57.

27. Gregory, "Pierre Charron's 'Scandalous Book,' " 98.

28. Ibid., 92.

29. Kristeller, *Renaissance Thought and its Sources.*

30. Bouwsma, "Changing Assumptions in Later Renaissance Culture," 76.

31. Kelley, *Foundations of Modern Historical Scholarship*; Struever, *The Language of History in the Renaissance.*

32. This document purported to validate the Catholic Church's claim to temporal power on the grounds that Constantine had bequeathed the Italian portion of his Empire to the Church. From analyzing the Latin text, Valla realized that the document was a forgery, probably written in the eighth century.

33. Waswo, *Language and Meaning in the Renaissance.*

34. Ong, *Rhetoric, Romance, and Technology*, 65.

35. Woodhouse, *Baldesar Catiglione*; Montrose, "Shaping Fantasies and Figurations of Gender and Power in Elizabethan Culture"; Greenblatt, *Renaissance Self-Fashioning*; Burke, *Fortunes of the Courtier*, 32.

36. Waswo, *Language and Meaning*, 219.

37. Ibid., 131.

38. Ibid., 132.

39. Grafton, *New Worlds, Ancient Texts*, 5ff.

40. Miller, "Making the Paris Polyglot Bible"; idem, "The 'Antiquarianization' of Biblical Scholarship."

41. Harrison, *The Bible, Protestantism, and the Rise of Natural Science*, 117.

42. Ibid., 114.

43. Shuger, *The Renaissance Bible*, 47.

44. Ibid., 24.

45. Simon, *A Critical History of the Old Testament*; Spinoza, *Tractatus Theological-Politicus.*

46. Shuger, *The Renaissance Bible*, 21.

47. Ex 28:30; Lev 8:8; Deut 33:8; Nu 27:21; 1 Sam 14:37–41.

48. Shuger, *The Renaissance Bible*, 29. Momigliano, *The Classical Foundations of Modern Historiography*; Miller, *Momigliano and Antiquarianism.*

49. Translations of the Hebrew Bible into Aramaic (250 BCE–300 CE).
50. A collection of rabbinic laws (c. 200 CE).
51. Commentaries on the Midrash (200–550 CE).
52. Shuger, *The Renaissance Bible*, 33.
53. Grafton, "Rhetoric, Philology and Egyptomania," 189; idem, *Joseph Scaliger*.
54. Hendrix, "Martin Luther and Albrecht of Mainz."
55. The word "fact" only began to acquire its modern meaning as something that is indubitably true or real in the late sixteenth century. Debrock, "Aristotle, Wittgenstein, *alias* Isaac Newton between Fact and Substance."
56. Harrison, *The Bible, Protestantism, and the Rise of Natural Science*, 98.
57. Shuger, *The Renaissance Bible*, 53.
58. Hayden, *The Counter-Renaissance*, 143.
59. William of Ockham (c. 1288–1348), an English Franciscan friar and scholastic philosopher, was a leading nominalist, arguing that only individuals exist and that universal forms or essences were concepts with no actual reality.
60. Certeau, *The Mystic Fable*, ch. 2.
61. Alpers, *The Art of Describing*, 85.
62. Ibid., 163.

CHAPTER 2

1. My discussion of Morandi is based on Brendan Dooley's fascinating book *Morandi's Last Prophecy and the End of Renaissance Politics*. The Roman "rumor mill" is mentioned on p. 1.
2. Ibid., 3.
3. Ibid., 113.
4. Ibid.
5. Dooley, *Morandi's Last Prophecy*, 146.
6. Geertz, "An Anthropology of Religion and Magic"; Ehnmaark, "Religion and Magic—Frazer, Söderblom, and Hägerström"; Pettersson, "Magic—Religion."
7. Styers, *Making Magic*, 4.
8. Frazer, *The Golden Bough*, 13.
9. Styers, *Making Magic*, 221.
10. Cf. Wittkower and Wittkower, *Born Under Saturn*.
11. Taylor, *Sources of the Self*, 192; Glucklich, *The End of Magic*.
12. Roper, *The Holy Household*.
13. Benavides, "Magic, Religion, Materiality."
14. Cf. Said, *Culture and Imperialism*.
15. Geertz, "An Anthropology of Religion and Magic." Thomas replied to Geertz in "An Anthropology of Religion and Magic, II." Kieckhefer also argues that for medieval people magic was not only real but rational ("The Specific Rationality of Medieval Magic").
16. Borchardt, "The Magus as Renaissance Man."

17. Luther, "Lectures on Galatians," *Works*, 26, 190. Oberman, *Luther: Man between God and the Devil.*

18. Jacob and Wilhelm Grimm, Doctor Luther zu Wartburg, *Deutsche Sagen*, vol. 2 (1818), no. 562. Translated by D. L. Ashliman (http://www.pitt.edu/~dash/legend562.html).

19. Calvin, *Institutes of the Christian Religion*, Bk 1, ch. xiv, 173.

20. Strauss, *Luther's House of Learning*; Cressy in *Bonfires and Bells*; Hutton in *The Rise and Fall of Merry England.* These books describe the way Protestant reformers challenged and adapted rituals of the pre-Reformation Church, but with only partial success. Cf. Scribner, "The Reformation, Popular Magic, and the 'Disenchantment of the World' "; Walsham, "The Reformation and 'The Disenchantment of the World' Reassessed."

21. Cameron, *Enchanted Europe*, 158.

22. Saler, "Modernity and Enchantment."

23. English Puritans wore their hair closely cropped unlike the Royalists, who preferred long hair in ringlets. Hence the term "round heads" for Puritans.

24. Heinrich Grimm, "Die deutschen 'Teufelbucher' des 16 Jahrhuunderts:"; Osborn, *Die Teufelliteratur des XVI Jahrhunders*; Roos, *The Devil in Sixteenth-Century German Literature*. Roos gives the much higher estimate of 240,000 devil books, but this includes the tracts that were bound together in the three editions of Sigmund Feyerabend's *Theatrum Diabolorum* (1569, 1575, 1587–88).

25. Roos, *The Devil in German Literature*, 48.

26. Stark, *For the Glory of God.*

27. Styers, *Making Magic*; Park, "Medicine and Magic," 132; Jolly, "Medieval Magic," 25.

28. Stephens, *Demon Lovers.*

29. Walker, *Spiritual and Demonic Magic.*

30. Clark, "Witchcraft and Magic," 110.

31. Wilson, *The Magical Universe.*

32. Jolly, "Medieval Magic," 55.

33. Walker, *Spiritual and Demonic Magic*; Zika, "Reuchlin's *De Verbo Mirifico*."

34. Walker, *Spiritual and Demonic Magic*, chs. 1–2.

35. Ficino, *Opera Omnia*, II: 1320.

36. Walker, *Spiritual and Demonic Magic*, 10.

37. Maxwell-Stuart, *The Occult in Early Modern Europe*, 116.

38. Ibid., 116.

39. Eamon, "Technology as Magic in the Late Middle Ages and the Renaissance."

40. Lehrich, *The Language of Demons and Angels*, 37.

41. Clark, "Witchcraft and Magic in Early Modern Culture," 150.

42. Lehrich, *The Language of Demons*, 197.

43. *Henrie Cornelius Agrippa of the Vanities and Uncertainties of Artes and Science*. Englished by Ja[mes] San[ford] gent. (1569), To the Reader, sig iii.

44. Clark, "Witchcraft and Magic in Early Modern Culture," 163.

45. Pingree, ed., *Picatrix*.

46. Moran, ed., *Patronage and Institutions*; idem, *The Alchemical World of the German Court*; Smith, *The Business of Alchemy*.

47. Trevor-Roper, *The European Witchcraze*.

48. Monter, "Witch Trials in Continental Europe."

49. Stephens, *Demon Lovers*, 366.

CHAPTER 3

1. Soergel, "Miracles," 64–66; idem, *Wondrous in His Saints*.

2. Strauss, "Success and Failure in the German Reformation," 59, 61–62.

3. Soergel, "Miracles," 64.

4. Scribner, "Incombustible Luther."

5. Soergel, *Wondrous in His Saints*; Strauss, *The German Single-Leaf Woodcut*; Friedman, *The Battle of the Frogs and Fairford's Flies*.

6. Soergel, "Miracle, Magic, and Disenchantment in Early Modern Germany," 231.

7. Schutte, " 'Such Monstrous Births,' " 97.

8. *The Diary of Ralph Josselin, 1616–1683*. The bee sting occurred on 5 September 1644. The death of Mary is described in the journal entries for May 22–27, 1650.

9. Cited in Maxwell-Stuart, "Rational superstition . . . ," 186, n12.

10. Samuel Sewell, *Diary* (October 1, 1697).

11. Soergel, "Miracle, Magic, and Disenchantment in Early Modern Germany," 232.

12. Walsham, *Providence in Early Modern England*.

13. Shaw, *Miracles in Enlightenment England*, 97.

14. *Fox's Book of Miracles*.

15. Clarkson, *Death, Disease & Famine*, 105.

16. Shaw, *Miracles in Enlightenment England*.

17. See, for example, *An Exact Relation of the Wonderful Cure of Mary Maillard* (London, 1730).

18. Shaw, *Miracles*, 119–20.

19. Ibid., 128.

20. Ibid., 128–129.

21. "Enthusiasm" became a topic of great concern in the late seventeenth and eighteenth centuries. Cf. Heyd, "*Be Sober and Reasonable*."

22. Shaw, *Miracles*, 3.

23. Ibid., 76.

24. Nicolson, ed. *Conway Letters*, 273.

25. Shaw, *Miracles*, 6.

26. Ibid., 85.

27. Ibid., 88.

28. Stewart, *The Rise of Public Science*, xxiii.

29. Hysteria was derived from the Greek word for uterus and therefore connected with women and by implication mothers. Plato's dialogue *Timaeus* describes the uterus wandering around a woman's body, strangling her as it reaches the chest and thereby causing disease.

30. Stubbe, *The Miraculous Conformist*, 4. In calling Greatrakes a "conformist," Stubbe was making it clear that he was not a divinely inspired "Enthusiast."

31. "Wen," a cyst containing sebaceous matter.

32. Stubbe, *The Miraclulous Conformist*, 6–7.

33. Shaw, *Miracles*, 83.

34. Ibid., 84.

35. Walker, *Unclean Spirits*.

36. Shaw, *Miracles*, 86.

37. Stubbe, *The Miraculous Conformist*, 14; Shaw, *Miracles*, 86; Kaplan, "Greatrakes, the Stoker."

38. Shaw, *Miracles*, 96.

39. Harrison, *The Bible, Protestantism, and the Rise of Natural Science*, 146.

40. Ibid., 146.

41. Hoppen, "The Nature of the Early Royal Society"; Dear, "Miracles, Experiments and the Ordinary Course of Nature."

42. Glanvill, *Some Philosophical Considerations Touching the Being of Witches and Witchcraft*, 4.

43. Hunter, *The Occult Laboratory*, 30.

44. Ibid., 82.

45. Ginzburg, *Ecstasies*.

46. Coudert, *The Impact of the Kabbalah*, ch. 6.

47. Hunter, *The Occult Laboratory*, 97.

48. Ibid., 170.

49. Ibid., 91.

50. Ibid., 86.

51. Ibid., 86–87, 150.

52. Ibid., 94.

53. Ibid., 87.

CHAPTER 4

1. Kittredge, *Witchcraft in Old and New England*, 335. He continues, "No English work on the subject had a more powerful influence."

2. Glanvill, *Saducismus triumphatus*, 169–71.

3. Cody, "The Doctor's in Labour"; Todd, "Three Characters in Hogarth's Cunicularii." Cf. Paulson, *Hogarth's Harlot*, 75–76.

4. *Disquisitionum Magicarum* (1606), Bk. 5, sect. 4, no. 28.

5. Over the past fifty years estimates of the number of people accused and executed for witchcraft has been vastly reduced from an initial figure of nine

million to approximately 35,000, 29,000 of whom were women. Cf. Monter, "Witch Trials in Continental Europe."

6. Boguet, *A discourse on witchcraft*. "There are witches by the thousands everywhere multiplying like worms in a garden," dedication, xxxiv.

7. Trevor-Roper, *The European Witch-Craze*, 80.

8. Monter, "Inflation and Witchcraft," 371–89.

9. Zika, *Exorcising our Demons*, 269; Ankarloo and Clark, ed., *Witchcraft and Magic in Europe*. The *Period of the Witch Trials*, introduction, especially ix.

10. Monter, "Witch Trials in Continental Europe."

11. Barstow, *Witchcraze*.

12. Laqueur, *Making Sex;* Schiebinger, *The Mind Has No Sex*.

13. Stollberg, "A Woman Down to Her Bones."

14. The topic of witchcraft and demonology comes up in all kinds of books because both were linked to the scientific, historical, religious, and political debates of the period. Cf. Clark, *Thinking with Demons*, viiiff.

15. Ibid., 111.

16. Ibid., 129. Cf. Clark, "The 'Gendering' of Witchcraft in French Demonology," 432.

17. Ibid. Cf. Wilson, "Contraries in Sixteenth Century Scientific Writing"; Davis, *Society and Culture in Early Modern France*, ch. 5.

18. Clark, "The 'Gendering' of Witchcraft," 427.

19. Brown, "Sorcery, Demons, and the Rise of Christianity."

20. Karlsen, *The Devil in the Shape of a Woman*, 51–52.

21. In *Witch-hunting in South Western Germany*, Midlefort attributes the decline of witch panics to the disinclination of male jurors and male judges to convict male suspects. A similar situation occurred in Salem. Cf. Boyer and Nissenbaum, *Salem Possessed*, 32–33.

22. Demos, *Entertaining Satan*, 64–70; idem, "Underlying Themes."

23. Karlsen, *The Devil in the Shape of a Woman*, 73ff, 295n93; Midelfort, *Witch-hunting in South Western Germany*, 184ff.

24. Mcfarlane, *Witchcraft in Tudor and Stuart England*, 160.

25. Midlefort denied the implications of his own research, claiming that there was no particular tradition of misogyny in the sixteenth century. He later changed his position in "Heartland of the Witchcraze." Others agree. Cf. Monter, *Witchcraft in France and Switzerland*, 196–98, and Larner, *Enemies of God*, 92, 197, although the latter waffles. Cf. Barstow, *Witchcraze*, 7.

26. Scott, "Gender: A Useful Category of Historical Analysis."

27. Kelso, *Doctrine for the Lady of the Renaissance*, 6, 10–13; Kunzle, *The Early Comic Strip*, 222–57: Klaits, *Servants of Satan*, 52.

28. Thomas, "The Double Standard"; idem, "Puritans and Adultery"; Schnucker, "La position puritaine à l'egard de l'adulte"; Monter, "Women in Calvinist Geneva."

29. Roper emphasizes the drastic change involved in the Protestant rejection of celibacy. She thinks the "crisis" in gender relations produced by the Reformation was never resolved (*Oedipus & the Devil*, 38).

30. Karlsen, *Devil in the Shape of a Woman*, 149.

31. Cooper, *The Mystery of Witchcraft*, 206.

32. Brauner, "Luther on Witchcraft."

33. For the increasing intolerance of sexual deviancy in relation to the witch hunts, see Klaits, *Servants of Satan*; Roper, *Oedipus & the Devil*; Quaife, *Wanton Wenches and Wayward Wives*; Sharpe, "The History of Crime in Late Medieval and Early Modern England"; Zika, "Fears of Flying."

34. Monter, "Pedestal and Stake,"132.

35. Thomas, "Women and the Civil War Sects"; Cross, "Great Reasoners in Scripture"; Nadelhaft, "The Englishwoman's Sexual Civil War"; Karant-Nunn, "Continuity and Change."

36. Leonard, "Female Religious Orders."

37. Battis, *Saints and Sectarians*, 242.

38. Hall, ed., *The Antinomian Controversy*, 382–83.

39. Costa, "Spanish Women in the Reformation," 99.

40. Ariès and Duby, ed., *A History of Private Life*, 60.

41. Ibid., 61.

42. Zika discusses the change to sexually charged images of naked witches (*Exorcising our Demons*, ch. 6).

43. Zika, "Dürer's Witch"; Mellinkoff, "Riding Backwards"; Underdown, "The Taming of the Scold"; Moxey, *Peasants, Warriors, and Wives*; Wiltenburg, *Disorderly Women and Female Power*; Dresen-Coenders, ed., *Saints and She-devils*; Clark, *Thinking with Demons*, ch. 2.

44. While Aristotle was acting as Alexander the Great's tutor, he warned Alexander against his favorite courtesan Phyllis on the grounds that women were the downfall of many strong men. To get her revenge, Phyllis made the philosopher fall in love with her and demanded that as a sign of his love he allow her to ride on his back as on a horse. Alexander witnessed this scene and concluded that women could never to be trusted if even old and venerated philosophers were powerless against them. Cf. Hutchinson, "The Housebook Master and the Folly of the Wise Man"; Sarton, "Aristotle and Phyllis."

45. Scribner, "Ways of Seeing in the Age of Dürer"; Russell, *Eva/Ave*.

46. Zika, credits Baldung with establishing a "new iconography" of witchcraft that "inextricably linked" witchcraft to women's bodies and women's labor and connected all three to inversion (Zika, "Baldung [Grien] Hans"). Cf. Zika, "She-Man"; Hoak, "Witch Hunting and Women"; Shestack, "An Introduction to Hans Baldung"; Charles W. Talbot, "Baldung and the Female Nude"; Hults, "Baldung and the Reformation"; Hults, "Baldung and the Witches of Freiburg"; Koch, *Hans Baldung Grien*.

47. Midelfort, "Heartland of the Witchcraze," 21; Monter, "Pedestal and Stake," 130; Levack, *Witch-Hunts in Early Modern Europe*, 19.

48. Kunzle, *Early Comic Strip*, 226.

49. Ibid., 229.

50. Coupe, *The German Illustrated Broadsheet*, 51.

51. Flandrin, *Families in Former Times*.

52. Gardner, *The Life and Times of Chaucer*, 53.
53. Warnicke, *Women of Renaissance and Reformation*, 156.
54. Karant-Nunn, "Continuity and Change"; Monter, "Women in Geneva."
55. Coupe, *German Broadsheet*, 2:145ff; Chaucer, "The Merchant's Tale."
56. Roper describes the situation in sixteenth century Augsburg: " . . . The Council regularly put far more detailed sheets of questions to women than to men, demanding to be told the full details of women's sexual encounters" (*The Holy Household*, 85).
57. Ibid., 85. In his *Anatomy of Melancholy* (1621), Robert Burton proclaims, "Of women's unnatural, insatiable lust, what country, what village doth not complain."
58. Medick, "Village Spinning Bees."
59. In *Solitary Sex*, Lacquer argues that concern with masturbation only began in the eighteenth century, but according to Roper, it was "one of the most abhorred sins" much earlier (*The Holy Household*, 65).
60. Broedel, *The Malleus Maleficarum*. Some scholars think that Kramer was the actual author, but that he added Sprenger's name to give his work greater authority.
61. Sidney Anglo, "Evident Authority and Authoritative Evidence."
62. Kramer and Sprenger, *Malleus Maleficarum*, 152–53; Smith, "The Flying Phallus and the Laughing Inquisitor"; Stephens, "Witches who Steal Penises."
63. Stephens, *Demon Lovers*, 366. Clark makes a similar point in *Thinking with Demons*, 29.
64. Stephens, *Demon Lovers*, 19.
65. Ibid., 32.
66. Ibid., 37
67. Ibid., 40.
68. Broedel, *The Malleus Maleficarum*, 6–7.
69. Ibid., 94.
70. Ibid., 176.
71. Stephens, *Demon Lovers*, 35.
72. Sextus Empiricus (c. 160–210 CE), a physician and philosopher primarily known for his skeptical views about the possibility of all kinds of knowledge.
73. Stephens, *Demon Lovers*, 235.
74. Ibid., 174.
75. Shapin and Schaffer, *Leviathan and the Air-Pump*, 104–5. Dear also discusses how members of the Royal Society tried to establish "facts" by writing ostensibly objective "reports."
76. Cohn, *The Pursuit of the Millennium*, 270.
77. Broedel, The *Malleus Maleficarum*, 179.
78. Wiesner, *Women in Early Modern Europe*.

CHAPTER 5

1. Schilling, "Confessional Europe"; idem, "Confessionalisation and the Rise of Religious and Cultural Frontiers"; Reinhard, "Gegen reformationals

Modernisierung?"; idem, "Reformation, Counter-Reformation and the Early Modern State"; Johnston, "The Reformation and Popular Culture."

2. Briggs, "The Science of Sin," 23.

3. Thomas, *Religion and the Decline of Magic*, 444.

4. Elias, *The Civilizing Process*. For a recent investigation of Elias's thesis, see Burke, Harrison, and Slack, ed., *Civil Histories*.

5. Roper, *Oedipus & the Devil*, 148.

6. Douglas, *Purity and Danger*, 4.

7. Bouwsma, *John Calvin*, 34.

8. Ibid.

9. Ibid., *Calvin*, 39.

10. Ibid.

11. Ibid.

12. Ibid., 46.

13. Ibid., 36.

14. Elias, *The Civilizing Process*; Douglas, *Purity and Danger*.

15. Douglas, *Purity and Danger*, 54.

16. On the themes of cannibalism incorporation, and colonialism in the early modern period, see Obeyesekere, *Cannibal Talk*; Avramescu, *An Intellectual History of Cannibalism*; and Kilgour, *From Communion to Cannibalism*.

17. Chatellier, *The Europe of the Devout*, 163.

18. Ibid., 66.

19. Ibid., 67.

20. Ibid., 69.

21. Ariès and Duby, ed., *A History of Private Life*, 80.

22. Chatellier, *The Europe of the Devout*, 42.

23. Ibid., 44.

24. Chatellier, *The Europe of the Devout*, 163.

25. Certeau, *The Mystic Fable*, 88.

26. Laporte, *History of Shit*, xi.

27. Molière, *Tartuffe*, Act III, scene ii.

28. Coudert, "Sewers, Cesspools, and Privies."

29. Diary of Cotton Mather, 1681–1708. "Massachusetts Historical Society Collections," 357.

30. Greven, *The Protestant Temperament*, 67.

31. Porter, *Flesh in the Age of Reason*, 25; Cockaye, *Hubbub*.

32. Correll, *The End of Conduct*, 19–20.

33. Horowitz, "Aristotle and Women"; McLaughlin, "Equality of Souls, Inequality of Sexes."

34. Bourdieu, "Structures and the Habitus."

35. Correll, *The End of Conduct*, 130.

36. Eby, *Early Protestant Educators*, 22–24. "School Ordinance from the Würtemberg Church Ordinance, 1559": ". . . And inasmuch as in some German schools not only the boys, but also the little girls are sent to school, we determine

that in such schools the children be separated, the boys alone and the little girls also be separately placed and taught, and that the schoolmaster by no means allow them to run back and forth among each other, or to have any disorderly relations with each other or to sleep together."

37. Wiesner, *Women and Gender*, 252ff.; Schilling, "Fruheneuzeitliche Formierung und Disziplinierung von Ehe, Familie und Erziehung." Leonard discusses the fear both Catholic and Protestant authorities had of female freedom and activities outside of the cloister or marriage in "Female Religious Orders."

38. Karant-Nunn, "Reformation Society, Women and the Family."

39. Roper, "Luther: Sex, Marriage and Motherhood," 33.

40. Walzer, *The Revolution of the Saints*, 28ff.

41. Ibid., 49.

42. Reinburg, "Hearing Lay People's Prayer," 29.

43. Roper, *The Holy Household*; Morgan agrees (Cf. *The Puritan Family*, 19). For a dissenting view, see Shuger's chapter on "Nursing Fathers" in her *Habits of Thought in the English Renaissance*.

44. Roper, *The Holy Household*, 15. What is especially interesting is that this same radical reversal occurred in early Christianity and for much the same reasons. Cf. Elaine Pagels, *Adam, Eve, & the Serpent*; eadem, *The Gnostic Gospels*.

45. Stone, *Family, Sex, and Marriage*, 7. Davis quotes the preamble to a French ordinance for strengthening paternal power with the family in *Society and Culture in Early Modern France*, 128.

46. *Institutes of the Christian Religion*, II, 8, 252.

47. Eby, *Early Protestant Educators*, "Catechism of the Church of Geneva Being a form of Instruction for Children in the Doctrine of Christ (1541). Supervision by the Elders."

48. Roper, *The Holy Household*, 112.

49. Coudert, "Educating Girls."

50. Lectures on Genesis, Gen. 3:16, *Luther's Works*, I: 202 3. Cf. Sigrid Brauner, *Fearless Wives and Frightened Shrews*.

51. Cited in Roper, *Oedipus and the Devil*, 19.

52. Heseler, *Andreas Versalius' First Public Anatomy at Bologna 1540: An Eyewitness Report*, 181.

53. Stollberg "A Woman Down to Her Bones."

54. Wiesner, *Women and Gender in Early Modern Europe*, 241.

55. Luther, *Works* 47: 45.

56. Reinburg, "Hearing Lay People's Prayer," 31.

57. Orgel, "Prospero's Wife," 59.

58. Filmer, *Patriarcha and Other Political Writings*, 188.

59. Hobbes, *Leviathan*, ch. 20, par. 15.

60. Montrose, "A Midsummer Night's Dream"; Goldberg, "Fatherly Authority"; Orgel, "Prospero's Wife.

61. Montrose, "A Midsummer Night's Dream." Cf. Schücking, *The Puritan Family*, 87.

62. Montrose, "A Midsummer Night's Dream," 85ff.
63. Goldberg, "Fatherly Authority," 85ff.
64. Vanhaelen, *Comic Print and Theatre*, 129. Smith, *Masks of Wedlock*.
65. Ozment, *When Fathers Ruled*, 54, 202 n12.
66. Cited in Newman, "Alchemy, Domination, and Gender," 219.
67. Ibid., 224–25 n12.
68. Ibid. 219–20.
69. Jung, *Mysterium Coniunctionis*.
70. *Lambsprinck nobilis Germani philosophi antiqui libellus e [de] lapide philosophico e germanico versus latine redditus per Nicolaum Bernaudeum Delphinatem Medium* . . . (Francofurti, 1625). This was also published in Lazarus Zetznerus, *Theatrum chemicum*, vol. 3 (1613) and in *Musaeum hermeticum reformatum* (1678), which was translated into English by A. E. Waite as *The Hermetic Museum* (1893).
71. Maier, *Atalanta Fugiens*, 115.
72. Ibid., 205.
73. Keller, *Reflections on Gender and Science*, 53–54. Merchant takes a similar view of alchemy in *The Death of Nature*. Recent scholarship has shown that this contrast is no longer tenable. Cf. Allen and Hubbs, "Outrunning Atalanta: Feminine Destiny in Alchemical Transmutation"; Tiles, "Mathesis and the Masculine Birth of Time"; Newman, "Alchemy, Domination, and Gender"; Pesic, "Westling with Proteus."
74. Needham, *A History of Embryology*.
75. Caspar Bauhin, *De Hermaphroditorum Monstrosorumque Partuum Natura* (Oppenheim: Hieronymus Galler/Joahnn Theodor de Bry, 1614).
76. Jung, *Mysterium Conjunctionis*; idem, *Alchemical Studies*; Eliade, *Forgerons et alchemists*; Wind, *Pagan Mysteries in the Renaissance*; Merchant, *The Death of Nature*; eadem, *Earthcare*; Fox Keller, *Reflections on Gender and Science*.
77. Waddington, "The Bisexual Portrait of Francis I."
78. Wiesner points out that when a woman succeeded in domains reserved for men (literature, arts, sciences) she was said to have "overcome the limitations of her sex" or she was judged "a hermaphrodite." In other words, her accomplishments spring from some addition of maleness (*Women and Gender in Early Modern Europe*, 147–48). The journal *Diogenes* devotes its November 2005 issue to the subject of androgyny.
79. Daston and Park, "Hermaphrodites in Renaissance France."
80. Taufer, "The Only Good Amazon Is a Converted Amazon," 35–52; Woods, "Amazonian Tyranny"; Sullivan, "Amazons and Aristocrats."
81. Horowitz, introduction, *Playing with Gender*, ix.
82. Ibid., x.
83. Tyndale, "Exposition of Matthew V, VI, VII," *The Works of William Tyndale*, ed. G. E. Duffield (Philadelphia: Fortress Press, 1965), 238–39.
84. Haller and Haller, "The Puritan Art of Love."
85. William Gouge, *Domesticall Duties* (London, 1622).
86. John Brinsley, *A Looking-Glasse for Good Women* (London, 1645), 39.

87. Cited in Bouwsma, *Calvin*, 52–53.
88. Thomas, "Women in Civil War Sects."
89. Bainton, *Women of the Reformation*, 55.
90. Underdown, "The Taming of the Scold"; Amussen, "Gender, Family and the Social Order."
91. The theme of inversion is treated in Davis, *Society and Culture in Early Modern France*, ch. 5; Babcock, ed., *The Reversible World*; Stallybrass, "The World Turned Upside Down."
92. Greenblatt, *Renaissance Self-fashioning*.
93. Woodhouse, *Baldesar Catiglione*, 54.
94. Correll, *The End of Conduct*, 76.
95. Desiderius Erasmus, "On Good Manners for Boys, *Works*, 25: 269–89.
96. Correll, *The End of Conduct*, 58. Kahn emphasizes the "emotional vulnerability" of men at the time in *Man's Estate: Masculine Identity in Shakespeare*. Cf. Foyster, *Manhood in Early Modern England*.
97. Moore, *The Maiden's Mirror: Reading Material of German Girls in the Sixteenth and Seventeenth Centuries*, 28.
98. Douglas, *Purity and Danger*, 143.
99. Roper, *Oedipus & The Devil*, 107.
100. Ibid., 111.
101. Burke, *Popular Culture*, 207–43.
102. Barker-Benfield, *The Culture of Sensibility*, 52.
103. "Molly houses" were taverns or private rooms where homosexual and transvestite men could meet.
104. Bahlman, *The Moral Revolution*.
105. Balibar and Laporte, *Le Français National*.
106. Davis, "Proverbial Wisdom and Popular Errors," *Society and Culture in Early Modern France*.
107. Stone, "Literacy and Education in England."
108. Henningsen, *The Witches' Advocate*; Contreras and Henningsen, "Forty-four Thousand Cases."
109. Chatellier, *The Europe of the Devout*, 46, 51.
110. Ibid., xi. O'Malley, "The Society of Jesus."
111. Chatellier, *The Europe of the Devout*, 44.
112. Ibid., 236ff.
113. Ibid., 141.
114. Hull, *Sexuality, State, and Civil Society in Germany*, 25.
115. Clark, "Inversion, Misrule, and the Meaning of Witchcraft."
116. Brauner, "Martin Luther on Witchcraft."

CHAPTER 6

1. McManners, *Death and the Enlightenment*, 61–62.
2. Troyansky, *Old Age*, 80.

3. Ibid., 86–87.
4. McManners, *Death and the Enlightenment*, 84.
5. Lindemann, *Health & Healing*, 349.
6. Ibid., 373.
7. Bödeker, "Menscheit, Humanität, Humanismus."
8. Leonard, "Female Religious Orders"; Cohen "Asylums for Women in Counter-Reformation Italy."
9. Lindberg, *The European Reformations*, 113, 126; Stephen, "Philanthropy and Empire"; McClure, *Coram's Children*.
10. Cassirer, *The Philosophy of the Enlightenment*, 141.
11. Porter, *Flesh in the Age of Reason*, 435.
12. Barker-Benfield, *The Culture of Sensibility*, 106.
13. Nearly 200 books and pamphlets dealing with education were published between 1714–89. Cf. Bridgman, "Introduction." *Education in the 18th Century*.
14. Leibniz, *Theodicy*, section 280.
15. Deleuze, *The Fold: Leibniz and the Baroque*.
16. Hansch, *Godofredi Guilielmi Leibnitii Principia Philosophiae More Geometrico Demonstrata* (Frankfurt & Leipzig, 1728), 135.
17. Barker-Benfield, *The Culture of Sensibility*, 35
18. Diderot, "Eloge de Richardson" in the *Journal étranger* (January 1762). http://graduate.engl.virginia.edu/enec981/dictionary/25diderotC1.html.
19. Darnton, "Readers Respond to Rousseau," 232.
20. Schama, *Citizens*, 160.
21. McKeon, *The Origins of the English Novel*.
22. *Spectator*, I: 10.
23. Ibid.
24. Porter, *Flesh in the Age of Reason*, 386–87.
25. Baridon, "Les concepts de nature humaine et de perfectibilité."
26. Mulsow, "Vernünftige Metempsychosis"; Benz, "Die Reinkarnationslehre"; Altmann, "Lessings Glaube an de Seelenwanderung,"; McCalla, "Palingenesie Philosophique."
27. Porter, *Flesh in the Age of Reason*, 138.
28. Barker-Benfield, *The Culture of Sensibility*, 138–39.
29. Ibid., 139.
30. Paulson, *Hogarth's Harlot*, 23.
31. Addison, "On Popular Superstitions," last paragraph. http://addison.classicauthors.net/essaytales/essaytales3.html.
32. Porter, *Flesh in the Age of Reason*, 127.
33. Ibid., 108.
34. Conway, *The Principles of the Most Ancient and Modern Philosophy*.
35. Patrides, "The Salvation of Satan."
36. Porter, *Flesh in the Age of Reason*, 285.
37. Seigel, *The Idea of the Self*, 108.
38. Reill, *The German Enlightenment*, 15.

39. Neiman, *Evil in Modern Thought*, 2–8, 4.
40. Ibid., 43–49.
41. Ibid., 5.
42. Shorto, *Descartes' Bones*.
43. Shapiro, *John Wilkins*, 199–200.
44. Leibniz, *Tagebuch, Geschichtliche Aufsätze und Gedichte Gottfried Wilhelm Leibniz*, 189–91.
45. Nicolson has found the sources for many of the experiments described by Swift. Nicolson, "Virtuoso." See her edition of Thomas Shadwell's *The Virtuoso*." Cf. Houghton, "The English Virtuoso in the Seventeenth Century."
46. Gaukroger, *The Emergence of a Scientific Culture*, 37.
47. Middleton, "What did Charles II Call the Fellows of the Royal Society?"; Syfret, "Some Early Reactions to the Royal Society."
48. Hill, *A Review of the Works of the Royal Society*, 11.
49. Rossi, *Francis Bacon*; Gaukroger, *Francis Bacon*; Harkness, *The Jewel House*.
50. Bacon, *The Collected Works of Francis Bacon*, 9: 39.
51. Ibid, 5: 400.
52. Ibid, 5: 401.
53. Zwierlein, "Medea's Revenge."
54. Achinstein, "How To Be a Progressive without Looking like One"; Real, "Tradition and Progress in Bacon's *Advancement of Learning*."
55. *The Philosophical Writings of Descartes*, I: 109.
56. Ibid., I: 143.
57. *The Works of Monsieur St. Evremond*, I, xli–xlii.
58. The classic account of the conflict between the so-called "Ancients" and "Moderns" is Richard F. Jones's *Ancients and Moderns*. For more modern evaluations, see Levine, *The Battle of the Books*.
59. Arbellot, "Les routes de France au XVIIIe Siècle"; idem, "Autour des routes de poste." Cf. Girard, "Le triomphe de la cuisinère bourgeoisie."
60. Bacon did not invent this analogy but drew upon John of Salisbury (who drew, in turn, upon Bernard of Chartres).
61. Petrarch, *Letters to Familiar Authors*, 151.
62. Preface to *The Great Instauration* (1620).
63. Locke, *Essay on Human Understanding*, Epistle to the Reader, 10.
64. *Lord Chesterfield's Letters*, 163.
65. Nicolás Monardes, *Joyfull Newes out of the Newe Founde Worlde* (1577), fol. 34v.
66. Shelton, "Cabinets of Transgression"; Hooper-Greenhil, *Museums and the Shaping of Knowledge*.
67. Israel, *Radical Enlightenment*, 142–51.
68. Lowthorpe, ed., *The Philosophical Transactions and Collections to the End of the Year 1700*, 3: 225–35. Iain Pears's evocative novel *An Instance of the Finger Post* gives a fascinating account of Lower's work on transfusion.
69. On the importance of seeing the heart in terms of a pump, see Moran, *Distilling Knowledge*, 99.

70. Porter, *Flesh in the Age of Reason*, 228.
71. Ibid., 228, 117.
72. Temkin, *The Falling Sickness*.
73. Porter, *Flesh in an Age of Reason*, 54.
74. Gay, "The Enlightenment as Medicine and as Cure."
75. McManners, *Death and the Enlightenment*, 6.
76. Marie Antoinette inaugurated these extravagant hairstyles or "poufs," which could be up to a yard high. They were formed from a wire padded with wool, cloth, horsehair, and gauze, interwoven with the woman's own hair. The elaborate construction was stiffened with pomade and dusted with powder, both of which attracted vermin, requiring fashionable ladies to carry long handled head-scratchers. Marie Antoinette wore a "pouf à l'inoculation" to publicize the fact that she had persuaded the King to be inoculated against smallpox.
77. Bennett, "Instruments and Illustrations."
78. Heilbron, "Domesticating Science in the Eighteenth Century," 10.
79. Tomas, "Naturalistic Illustrations and Collections in Tuscany."
80. Bennett, "Instruments and Illustrations," 151; Spary, *Utopia's Garden*; Jardine, Secord, Spary, ed., *The Cultures of Natural History*.
81. Jacob and Sturkenboom, "A Woman's Scientific Society in the West."
82. Ibid., 221.
83. Ibid.
84. James Thompson and John Conduitt described Newton as "a Saint and his discoveries might well pass for miracles" (Gaukroger,*The Emergence of a Scientific Culture*, 25). Iliffe, " 'Is He Like Other Men?' " Patricia Fara, *Newton*, 199: "In religious iconography, the infant's apple indicates that Christ, the second Adam, will redeem humanity. For Bacon's followers, Newton became the new Adam who would uncover God's mathematical laws of nature." Galileo has received similar treatment. A tooth, thumb, and index finger belonging to him are now on display at the Galileo Museum in Florence. In much the same way saints' relics are displayed in churches and shrines.

CHAPTER 7

1. For a concise discussion of historiographical debates about the Scientific Revolution, see Shapin, *The Scientific Revolution*, 1–8, 170–74.
2. Butterfield, *The Origins of Modern Science*, viii.
3. Westfall, "The Scientific Revolution Reasserted," 44.
4. Ibid., 50.
5. Henry, *The Scientific Revolution*, 17
6. Dear, *Revolutionizing the Sciences*.
7. Jacob, "The Truth of Newton's Science and the Truth of Science's History," 331. For a discussion of the Scientific Revolution and early modern science in general, see the excellent collection of articles in Park and Daston, ed., *The Cambridge History of Science*. Vol. 3: Early Modern Science. See also Applebaum, ed.,

Encyclopedia of the Scientific Revolution and Applebaum, ed., *The Scientific Revolution and the Foundations of Modern Science.*

8. Cohen, *Revolution in Science.*

9. Paul, *Science and Immortality.*

10. Jacob, "The Truth of Newton's Science," 319.

11. Cohen, *The Scientific Revolution*, 501.

12. Brooke, *Religion and Science*; Feldhay, "Religion"; Lindberg and Numbers, *God & Nature*; Pagel, "The Debt of Science and Medicine to a Devout Belief in God"; idem "Religious Motives in the Medical Biology of the XVIIth. Century"; Popkin, "The Religious Background of Seventeenth Century Philosophy"; Lindberg and Numbers, *When Science and Christianity Meet*; Brooke, Osler, and van der Meer, *Science in Theistic Contexts.*

13. Brooke, *Science and Religion*, 19ff., 95.

14. Klaaren, *Religious Origins of Modern Science.*

15. Gillispie, "Natural History, Natural Theology and Social Order."

16. Pagel, "The Debt of Science to a Devout Belief in God."

17. Kubrin, "Newton and the Cyclical Cosmos."

18. Cited in Harrison, *The Bible, Protestantism, and the Rise of Natural Science*, 174.

19. Ibid., 174.

20. Ibid., 172.

21. Clarke, *A Demonstration of the Being and Attributes of God*, 91.

22. Williamson, *Apocalypse Then: Prophecy and the Making of the Modern World*, 266.

23. Ibid., 1–2.

24. Coudert and Shoulson, *Hebraica Veritas.*

25. Coudert, *The Impact of the Kabbalah in the Seventeenth Century.*

26. The Kabbalah is the commonly used term for the mystical teachings of Judaism originating after the twelfth century. It was thought to represent the esoteric and unwritten aspect of the divine revelation granted to Moses on Sinai, while the Bible was the exoteric portion of the same revelation. In the sixteenth century a new form of Kabbalah was produced by the students of Isaac Luria (1534–72). While this built on earlier Kabbalistic thought, there was an emphasis on the doctrines of reincarnation and universal salvation. The literature on the Kabbalah is vast. The best place to start is with the work of Gershom Scholem and Moshe Idel.

27. Idel, *New Perspectives on the Kabbalah*; Coudert, "The Kabbalah, Science, and the Enlightenment."

28. Idel, *The Kabbalah*, 170.

29. Werblowsky, "Messianism in Jewish History," 48.

30. Bacon, *Philosophical Works*, 46.

31. Browne, *Pseudodoxia Epidemica*, I: 5.

32. Harrison, *The Bible, Protestantism, and the Rise of Natural Science*, 205.

33. *On First Principles*, iv, iii, 1.

34. Daneau, *The Wonderful Workmanship of the World*, fol. 7v.

35. Lemaius, *An Herbal for the Bible*, 6.

36. *The Alphabet of Nature*, Preface, 10r.

37. More, *Conjectura Cabbalistica* (1653). Blair, "Mosaic Physics and the Search for a Pious Natural Philosophy in the Late Renaissance."

38. This was first published in 1629. Bennett and Mandelbrote comment on the punning nature of the title with its play on the author's name ("Park-in-sun's park on earth") in *The Garden, the Ark, the Tower, the Temple*, 62.

39. Harrison, *The Bible, Protestantism, and the Rise of Natural Science*, 231–32.

40. Bennett and Mandelbrote, *The Garden, the Ark, the Tower, the Temple*, 53–54.

41. Ibid., 59–60.

42. Ibid., 59; Shapiro, *John Wilkins*.

43. Harrison, *The Bible, Protestantism, and the Rise of Natural Science*, 238.

44. Thomas, *Man and the Natural World*, 255; Harrison, *The Bible, Protestantism, and the Rise of Natural Science*, 242.

45. Ibid., 244–46.

46. Bennett and Mandelbrote, *The Garden, the Ark, the Tower, the Temple*, 73ff. Cf. Harrison, "Original Sin and the Problem of Knowledge in Early Modern Europe."

47. Bouwsma, *John Calvin*, 47.

48. Findlen, *Possessing Nature*.

49. Cf. Borst, *Der Turmbau von Babel*; Knowlson, *Universal Language Schemes in England and France*; Aarsleff, *From Locke to Saussure*; Bono, *From Ficino to Descartes*, vol. 1 of *The Word of God and the Languages of Man*; Ecco, *The Search for the Perfect Language*; Markley, *Fallen Language*; Rossi, *Logic and the Art of Memory*.

50. Cf. Demott, "Comenius and the Real Character in England"; Rossi, *Logic and the Art of Memory*. Salmon downplays Comenius's influence in "Language-planning in Seventeenth century England"; Knowlson agrees in *Universal Language Schemes*, 57, 74.

51. Turnbull, *Hartlib, Dury and Comenius*; Webster, *The Great Instauration*; idem, *Samuel Hartlib and the Advancement of Learning*.

52. *The Petty Papers*, 2:150.

53. Wilkins, *Essay Towards a Real Character and a Philosophic Language* (London, 1668), preface.

54. Couturat, *La Logique de Leibniz*.

55. Sprat, *History of the Royal Society*, 113.

56. Ibid., 63.

57. Lüthy, "The Confessionalization of Physics," 108.

58. Debus, *Science and Education in the Seventeenth Century*; Shapin and Schaffer, *The Leviathan and the Air Pump*.

59. De Vairasse, *The History of the Sevarites or Sevarambi*, 388. Cf. Paul Cornelius, *Languages in Seventeenth and Early Eighteenth Century Imaginary Voyages*.

60. On the esoteric roots of many language schemes, see Coudert, *The Impact of the Kabbalah*, chs. 4 & 5. Cf. F. M. van Helmont, *Alphabet of Nature*.

61. The Wisdom of Solomon 11:20: "You have arranged all things by measures and number and weight".

62. Rosenau, *Vision of the Temple*; van Pelt, *Tempel van de Wereld*; idem, "Through the Temple of Solomon to the Temple of Heaven."

63. Harrison, *The Bible, Protestantism, and the Rise of Natural Science*, 4.

64. Cunningham and William, "De-centering the 'Big Picture' "; Cunningham, "Getting the Game Right. Some Plain Words on the Identity and Invention of Science."

65. Schmitt, *History of Universities*; Feingold, ed., *History of Universities*.

66. Vickers, ed., *Occult and Scientific Mentalities in the Renaissance*.

67. Historians of science do not agree on this. Henry, for example, claims that "the mechanical philosophy marks a definite break with the past and sets the seal upon the Scientific Revolution" (*The Scientific Revolution*, 69). But he also accepts that a certain amount of vitalism continued to exist among British mechanical philosophers (75ff.). Cf. McGuire, "Neoplatonism and Active Principles"; Hutchinson, "What Happened to Occult Qualities."

68. Principe, *The Aspiring Adept*; Hunter, ed., *Boyle Reconsidered*.

69. Dobbs, "Newton as Final Cause and First Mover," 34n18.

70. D'Alembert, "Preliminary Discourse," 76.

CHAPTER 8

1. Hanegraaff and Pijnenburg, ed., *Hermes in the Academy*.

2. Lehrich, *The Language of Demons*, 36.

3. Stuckrad, *Locations of Knowledge*.

4. Benz, "La Kabbale chrétienne en Allemagne"; Coudert, Hutton, et al., *Judaeo-Christian Intellectual Culture*; Goodman, ed., *Neoplatonism in Jewish Thought*; eadem, *Jewish and Islamic Philosophy*.

5. Clulee, *John Dee's Natural Philosophy*; Harkness, *John Dee's Conversations with Angels*.

6. Dobbs, *The Janus Faces of Genius*.

7. Walker, *The Ancient Theology*.

8. Yates, "The Hermetic Tradition in Renaissance Science." Cf. Yates, *The Rosicrucian Enlightenment*.

9. Schmitt, "Towards a Reassessment of Renaissance Arisotelianism"; idem, "Reappraisals in Renaissance Science"; idem, *Aristotle and the Renaissance*, 3, 12; idem, *The Aristotelian Tradition and Renaissance Universities*.

10. Vickers, "Analogy and Identity"; idem, "Frances Yates and the Writing of History"; idem, "On the Function of Analogy in the Occult."

11. Vickers, *Occult and Scientific Mentalities in the Renaissance*, introduction, 6.

12. Ibid., 8.

13. Cf. Ross, "Occultism and Phlosophy"; Schaffer, "Occultism and Reason."

14. Hunter, *Establishing the New Science*.

15. Pagel, *Joan Baptista van Helmont*; idem, "Religious Motives in the Medical Biology of the XVIIth Century"; idem, *Paracelsus*; idem, *William Harvey's Biological Ideas*.

16. Popkin, *The Third Force in Seventeenth Century Thought*.

17. Westfall, "Newton and the Hermetic Tradition"; idem, "Newton and Alchemy."

18. Hunter, ed., *The Occult Laboratory*.

19. Principe, *The Aspiring Adept*, 115ff.

20. Johnson and Wolbarsht, "Mercury Poisoning"; Spargo and Pounds, "Newton's 'Derangement of the Intellect' "; Principe, *The Aspiring Adept*, 179.

21. Copenhaver, "The Occultist Tradition and its Critics," 491–92.

22. Coudert, *Leibniz and the Kabbalah*.

23. Stolzenberg points out that while Kircher believed in astrology, he wasn't enthusiastic about it and considered most kinds of prognostication and astral magic illicit.

24. Findlen, ed., *Athanasius Kircher*.

25. Rossi, *The Dark Abyss of Time*, 96.

26. Gatti, *Giordano Bruno*.

27. Schuchard, *Restoring the Temple of Vision*; Åkerman, *Queen Christina*; eadem, *Rose Cross over the Baltic*; Moran, "Privilege, Communication, and Chemiatry."

28. Copenhaver, "Hermes Trismegistus, Proculus and the Question of a Theory of Magic in the Renaissance"; idem, "Natural Magic, Hermetism, and Occultism"; McGuire "Neoplatonism and Active Principles."

29. Trinkhaus, *In our Image and Likeness*, 2: 476–508, 519–24.

30. Kleinberg, "De agone christiano."

31. Brown, *Augustine of Hippo*, 395. Pagels also emphasizes Augustine extreme pessimism in *Adam, Eve, and the Serpent*, 149.

32. Leff, *Medieval Thought*, 37–38.

33. *Theologia Platonica*, 2: 247. On Pico's view of man and the question of grace, see Waddington, "The Sun at the Center." Waddington rejects Trinkhaus's contention that Pico's image of man is "non-operative" and "extraordinarily passive" (Trinkhaus, 519). However, he argues that Pico and Ficino still believed in the necessity of grace (86).

34. Ibid, 2: 256–58.

35. *De occultaphilosophia*, III: 55, 570/644. Lehrich, *The Language of Demons and Angels*, 197.

36. As Wallis has said of Plotinus, "[he] lacks any sense of sin or of the need for redemption. . . . Our true self is eternally saved and all that is required is to wake up to this fact, a process requiring self-discipline, but perfectly within the soul's own power" (*Neoplatonism*, 90).

37. Blumenberg, *The Legitimacy of the Modern Age*.

38. Newman, "Technology and the Alchemical Debate in the Late Middle Ages."

39. The idea that art is superior to nature becomes a conventional theme in the writings of those promoting natural magic and science. See Levin, *The Myth of the Golden Age in the Renaissance*; Parry, *The Golden Age Restored*.

40. Ginzburg, "High and Low."

41. *Confessions*, Bk 10. 35. 54. Cf. Steiner, "The Faust Legend"; Eamon, *Science and the Secrets of Nature*, 59ff; Peters, " 'Libertas Inquirendi' and the Vitum Curiositatis."

42. *Confessions*, Bk 10: 35. 55.

43. Steiner, "The Faust Legend." On the changing fortunes of Faust, see Hinman, "The Apotheosis of Faust."

44. Giammatti, "Proteus Unbound."

45. Pico compares man to a chameleon and to Proteus in his "Oration." He describes himself as Protean.

46. Harrison, "Curiosity, Forbidden Knowledge, and the Reformation of Natural Philosophy"; Benedict, *Curiosity*.

47. Newman, "Technology and the Alchemical Debate"; idem, *Promethean Ambitions*; Eamon; Clulee, *John Dee's Natural Philosophy*.

48. See Mebane, *Renaissance Magic*, 210–11, n27.

49. Gorceix, *La Bible des Rose-Croix*, lxiii.

50. Shuler describes the way alchemists from different Christian denominations could square the practice of alchemy with their religious beliefs. But in doing so, I would argue, they changed the character of their respective religions. This seems to be particularly true of Protestantism. Protestant alchemists and mystics (the two often overlapped) greatly attenuated the doctrines of human depravity and predestination (Schuler, "Spiritual Alchemies of Seventeenth Century England").

51. Moran, *Distilling Knowledge*, 65. Marxists scholars in the 1930s and 1940s, such as Edgar Zilsel and Benjamin Farrington, stressed the role of artisans and craftsmen in the Scientific Revolution almost to the exclusion of the role of scholars and natural philosophers. Recent studies take a more balanced approach. See Rossi and Pamela Smith.

52. Moran, *Distilling Knowledge*, 42.

53. Smith, *The Business of Alchemy*; eadem, *The Body of the Artisan*.

54. Moran, *Distilling Knowledge*, 187. This is similar to the argument made by Elizabeth Eisenstein in *The Printing Revolution in Early Modern Europe*, namely that print culture didn't necessarily promote new ideas as much as it fostered the juxtaposition of conflicting ideas, which in and of itself promoted new ideas.

55. Merton, "Science, Technology and Society"; idem, "Puritanism, Pietism and Science"; Shapin, "Understanding the Merton Thesis"; Webster, *The Great Instauration*.

56. Bennett, "The Mechanics' Philosophy and the Mechanical Philosophy."

57. On Catholic science, see Harris, "Transposing the Merton Thesis"; Hellyer, *Catholic Physics*; Feingold, ed., *Jesuit Science*.

58. Brooke, *Science and Religion*, 101. For a clear and judicious discussion of the Merton thesis. see Brooke, chapter 3.

59. Cf. the articles by Mulligan cited in the bibliography. Shapiro, "Latitudinarianism and Science in Seventeenth-Century England."

60. Trevor-Roper, *The Crisis of the Seventeenth Century*.

61. Aarsleff, *From Locke to Saussure*, 106.

62. Brann, *Trithemius and Magical Theology*, 86.

63. Rossi, *Logic and the Art of Memory*; Bolzoni and Parzen, *The Gallery of Memory*.

64. Slaughter, *Universal Language and Scientific Taxonomy*, 7.

65. Bordo, *The Flight to Objectivity*. Far from viewing Descartes as a rationalist, some of his contemporaries criticized him as an enthusiast because of his rejection of authority and reliance on private sources of knowledge. Cf. Heyd, "*Be Sober and Reasonable*," ch. 4. Casaubon was one contemporary who forcefully made this point by placing Descartes in the same category as the Quakers and by describing his method as a kind of "Mystical Theologie" (Casaubon, *A Treatise Concerning Enthusiasm*, 172–73).

66. Pelagius (c. 354–c. 420/40) was an ascetic monk, who denied the doctrine of original sin as developed by Augustine. He was declared a heretic at the Council of Carthage in 418 CE. On the revival of Pelagianism, see Heer, *Die Dritte Kraft*.

67. Trevor-Roper, *The Crisis of the Seventeenth Century*, 232–3: Cf. Westfall, "The Rise of Science and the Decline of Orthodox Christianity"; Brooke and Maclean, ed., *Heterodoxy in Early Modern Science and Religion*.

CHAPTER 9

1. The term was coined by Herbert Butterfield in his book *The Whig Interpretation of History* (1931). It describes the tendency to see history as the progressive development of present-day ideas. In this kind of history only the "winners" are considered worthy of study.

2. Newman and Principle, *Alchemy Tried by the Fire*, 13.

3. Ibid., 298–99.

4. What became the phlogiston theory was first developed by Johann Joachim Becher in 1667, when he postulated that there was a fiery element in substances released during combustion. It wasn't until Lavoisier discovered oxygen that the true nature of combustion was understood. But as Newman and Principe demonstrate, J. B. van Helmont, Starkey, and Boyle provided the source for the quantitative methods that led to Lavoisier's discovery (Ibid.).

5. See the works listed under Allen G. Debus in the bibliography.

6. Cook, "Medicine."

7. Van Helmont, *Oriatrike*, 875.

8. Van Helmont, *Deliramenta Catarrhi*, 8.

9. Van Helmont, *Opuscula*, 1023.

10. Van Helmont, *Ternary of Paradoxes*, in *Opuscula*, 1098.

11. Goldammer, "Paracelsische Eschatologie."

12. Van Helmont, *Oriatrike*, 387.

13. The idea that diseases are the result of the invasion or penetration of external forces was not necessarily modern. This was an aspect of ancient and primitive worldviews and supported the concepts of witchcraft and demonism.

14. Van Helmont, *Oriatrike*, 1102.
15. Newman and Principe, *Alchemy*, 59ff., 293.
16. Partington, "Joan Baptista van Helmont."
17. Newman and Principle, *Alchemy*, 69.
18. Ibid., 29. On van Helmont's indebtedness to Aristotle, see Pagel, *Joan Baptista van Helmont*, 10.
19. Van Helmont, *Oriatrike*, 536.
20. Ibid., 552.
21. Ibid., 582.
22. Cf. Walker, *Spiritual and Demonic Magic*.
23. Van Helmont, *Ternary of Paradoxes*, 58
24. Ibid., 58–59.
25. Ibid., 87.
26. Ibid., 59–60.
27. Walker, "Francis Bacon and Spiritus"; idem, "Medical Spirits in Philosophy and Theology from Ficino to Newton."
28. Koyré, *La Philosophie* de *Jacob Boehme*, 113.
29. Plotinus is neither consistent nor clear about the exact relationship between spirit and matter. Indeed, he is not clear about whether or not matter truly exists. Cf. *Enneads* I.8.
30. Newman, "The Corpuscular Theory of J. B. van Helmont.'
31. Van Helmont, *Ternary of Paradoxes*, 26–27; Broeckx, "Interrogatoires," 18.
32. Van Helmont, *Ternary of Paradoxes*, 74–75. William Harvey's discussion of conception has striking similarities to van Helmont's. He thought the womb looked and acted very much like the brain and that the semen did not affect it physically but produced a non-physical impression, somewhat like a thought, which caused the womb to conceive. Pagel, *William Harvey's Biological Ideas*, 270ff.
33. Broeckx, "Interrogatoires," 15–16, 43.
34. Debus, *The Chemical Philosophy*.
35. Sobel, *Longitude*, 41–42.
36. Van Helmont's run in with the Spanish Inquisition is discussed in Mévergnies, *Jean-Baptiste van Helmont*, 123–30, as well as in Broeckx, "Interrogatoires."
37. Walker, *Spiritual and Demonic Magic*.
38. Schmitt, *John Case and Aristotelianism in Renaissance*.
39. Van Helmont, "Logica inutilis," *Oriatrike*, 18. Bacon said much the same thing in *Novum Organum*, I, ii.
40. Rattansi, "Paracelsus and the Puritan Revolution."
41. Ibid.
42. Ibid., 22.
43. Ibid., 25.
44. Van Helmont, *Oriatriake*, 314.
45. Ibid., 23.
46. Newman and Principe, *Alchemy*, 58, 202.

47. , Ibid., 111, 163ff.

48. Pagel, *Paracelsus*, 56.

49. Van Helmont, *Oriatrike*, 523.

50. Newman and Principe, 111ff., especially 117.

51. Halleux, "Theory and Experiment in the Early Writings of Johan Baptista van Helmont."

52. Van Helmont, *Ternary of Paradoxes*, 36.

53. Ibid., 65

54. Hunter, *The Occult Laboratory*.

55. Van Helmont, *Oriatrike*, 1135.

56. Pagel, "The Debt of Science and Medicine to a Devout Belief in God."

57. Partington, "Jan Baptista van Helmont."

58. Pagel, *Joan Baptista van Helmont*, 124, n1.

59. Van Helmont's attitude was paralleled by Francis Bacon. See Rossi, *Francis Bacon*.

60. Yates, *The French Academies*; Evans, *Rudolph II and his World* ; idem, "Learned Societies in Germany in the Seventeenth Century"; Moran, *The Alchemical World of the German Court*.

61. Winner, *The Basic Ideas of Occult Wisdom*. Faivre has characterized the esoteric tradition in Western culture in somewhat similar terms in "Renaissance Hermeticism and the Concept of Western Esotericism." Hanegraaff modifies this in "On the Construction of 'Esoteric Traditions,' " as does von Stuckrad in *Was ist Estoerik* and Versluis in "What is Esoteric?"

62. Coudert, "Henry More and Witchcraft."

63. Hutchison, "What Happened to Occult Qualities in the Scientific Revolution?" 239.

64. Schaffer, "Godly Men"; Henry, "Occult Qualities and the Experimental Philosophy."

65. Hunter and Wootton, ed. *Atheism in Early Modern Europe*.

66. Debus, *The French Paracelsians*, 161ff; Secret, "Palingenesis, Alchemy and Metempsychosis in Renaissance Medicine."

67. Darnton, *Mesmerism*, 14.

68. Sagan, *The Demon-Haunted World*.

69. Dunbar, *The Trouble with Science*, ch. 7.

70. Nicolson, ed., *The Conway Letters*, 87.

71. Newman and Principle argue that Boyle's "first exposure to the full range of laboratory chymistry occurred at the hands of Starkey" after they met in1651 (*Alchemy*, 207). Starkey's allegiance to van Helmont is shown by his published work: *Nature's Explication and Helmont's Vindication* (London, 1657), in which he rejected the Galenic therapy of bleeding and purgation in favor of treating the *archeus* with chemical medicines. Starkey dedicated his *Pyrotechny Asserted and Illustrated* of 1658 to Boyle in the hope of getting him to side publicly with Hemontians against the Galenists, but Boyle was unwilling to get involved in this debate just as he had been unwilling to get involved in the debate about Greatrakes.

72. Clericuzio, "A Redefinition of Boyle's Chemistry," 563.
73. Clericuzio, "From van Helmont to Boyle," 319.
74. Diary entry for July 1662. Cited in Newman and Principe, *Alchemy*, 274.
75. Walton, "Boyle and Newton on the Transmutation of Water and Air, from the Roots of Helmont's Tree."
76. Clericuzio, "From van Helmont to Boyle," 331.
77. Newman and Principle, *Alchemy*, ch. 5.
78. Clericuzio, "From van Helmont to Boyle," 306. Newman and Principle, *Alchemy*, 289.
79. *The Correspondence of Henry Oldenburg*, I:355.
80. Clericuzio, "From van Helmont to Boyle," 319.
81. Ibid., 322.
82. Ibid., 325.
83. Ibid., 330.
84. Simpson, *Zymologia pysica, or a Brief Philosophical Discourse of Fermentation* (London, 1675).
85. Clericuzio, "From van Helmont to Boyle," 329.
86. Ibid., 334.
87. Newman, "Experimental Corpuscular Theory in Aristotelian Alchemy."
88. Cited in Newman and Principe, *Alchemy*, 299.

EPILOGUE

1. Letter to Dr. Benjamin Waterhouse, June 26, 1822 (http://lachlan.bluehaze.com.au/lit/jeff17.htm)
2. Finke and Stark, *The Churching of America 1776–1990*.
3. Berger, ed., *The Desecularization of the World*; Swatos and Olson, ed., *The Secularization Debate*.
4. Wallace, *Religion*, 265.
5. Tertullian, *Prescription against Heretics*, ch. 7, http://www.newadvent.org/fathers/0311.htm.
6. Langmuir argues that the charge of host desecration emerged at this time as one way to allay skepticism by offering "proof" that blood appeared when Jews desecrated the host. Hence the bread had changed into the body and blood of Christ. See *Towards a Definition of Antisemitism*. Stephens discusses the growing skepticism among Christians from the thirteen century up to the period of the witch hunts in *Demon Lovers* and sees the witch hunts as one way of assuaging these fears. Broedel agrees.
7. Dunbar, *The Trouble with Science*.
8. Spiro, "Religion," 116.
9. Stark and Bainbridge, *A Theory of Religion*, 23.
10. Ibid., 318.
11. Euben, *Islamic Fundamentalism and the Limits of Modern Rationalism*.
12. Cf. Strenski, *Why Politics Can't Be Freed From Religion*.

13. Speech given by Stern when he received an award from the Leo Baeck Institute for his book *The Politics of Cultural Despair* (reported in *New York Times*, January 6, 2005).

14. Thomas, *New York Times*, October 8, 1989.

15. Horkheimer and Adorno, *Dialectic of Enlightenment*, 3.

16. Ibid., 5.

17. Hollinger points out the way postmodernists altered the meaning of "modernism" by applying it to enlightenment thinkers and not to the generation of 1890–1930 (Nietzsche, Proust, Conrad, Yeats, Mann, Joyce, Stravinsky, Picasso, Nolde, Klimt, William James), who were identified as "moderns" by Lionel Trilling and others. As Hollinger says: "The postmodernists virtually plundered the old modernist canon, appropriating the thinkers they like for postmodernism and declaring the rest to be lingering echoes of the Enlightenment" (Hollinger, "The Enlightenment and the Genealogy of Cultural Conflict," 11–12).

18. Foucault, *Discipline and Punish*; idem, "What is Critique"; Gray, *Enlightenment's Wake*; Keller, *Reflections on Gender and Science*; Merchant, *The Death of Nature*; Wokler, "The Enlightenment Project and its Critics"; Schmidt, ed. *What is the Enlightenment?*; Baker and Reill, *What's Left of the Enlightenment?*"

19. Birken, *Hitler as a Philosopher*.

20. Cassirer, *The Philosophy of the Enlightenment*, x.

21. Ibid, xi–xii.

22. Sheehan, *The Enlightenment Bible*; idem, "The Enlightenment, Religion, and the Enigma of Secularization."

23. Ibid., 1062.

24. Neugebauer-Wölk, ed. *Aufklärung und Esoterik*.

25. Chatellier rejects the idea of the de-Christianization of Europe and discusses instead its new vitality in many places in *The Europe of the Devout*, 218; Gierl, *Pietismus und Aufklärung*. Barnett conclusively shows that the idea of a deist movement and deist radicals intend on destroying religion is a myth (*The Enlightenment and Religion*, 13).

26. Terrall, "Metaphysics, Mathematics, and the Gendering of Science in Eighteenth-Century France," 246.

27. Castle, "The Culture of Travesty," 251–270, 251. Cf. Castle *The Female Thermometer*.

28. Frost, "The Pacific Ocean."

29. Reill, *The German Enlightenment*, 214–15.

30. Habermas, *New Conservatism*, 201.

31. Habermas, *Structural Transformation of the Public Sphere*.

Bibliography

Aarsleff, Hans. *From Locke to Saussure: Essays on the Study of Language and Intellectual History*. London: Athlone Press Ltd., 1982.

Achinstein, Sharon. "How To Be a Progressive without Looking like One: History and Knowledge in Bacon's *New Atlantis*," *Clio* 17 (1988): 249–64.

Agustine, *Confessions*. Translated by F. J. Sheed and edited by Michael P. Foley. 2nd Ed. Indianapolis/Cambridge: Hackett Publishing Company, Inc., 2006.

Åkerman, Susanna. *Queen Christiana of Sweden and Her Circle: The Transformation of a Seventeenth Century Philosophical Libertine*. Leiden: Brill, 1991.

Åkerman, Susanna. *Rose Cross over the Baltic: The Spread of Rosicrucianism in Northern Europe*. Leiden: Brill, 1998.

Allen, Don Cameron. "The Degeneration of Man and Renaissance Pessimism," *Studies in Philology* 35 (1938): 202–27.

Allen, Sally G. and Joanna Hubbs, "Outrunning Atalanta: Feminine Destiny in Alchemical Transmutation," *Signs* 6 (1980): 210–29.

Alpers, Svetlana. *The Art of Describing: Dutch Art in the Seventeenth Century*. Chicago: Chicago University Press, 1983.

Amussen, S. D. "Gender, Family and the Social Order, 1560–1725." *Order and Disorder in Early Modern England*. Ed. A. Fletcher and J. Stevenson. Cambridge: Cambridge University Press, 1985.

Anglo, Sidney. "Evident Authority and Authoritative Evidence: The *Malleus Maleficarum*." *The Damned Art: Essays in the Literature of Witchcraft*. London: Routledge & Kegan Paul, 1977, 1–31.

Ankarloo, Bengt and Stuart Clark, ed. *Witchcraft and Magic in Europe. The Period of the Witch Trials*. Philadelphia: University of Pennsylvania Press, 2002.

Applebaum, Wilbur, ed. *Encyclopedia of the Scientific Revolution: From Copernicus to Newton*. New York: Garland, 2000.

Applebaum, Wilbur, ed. *The Scientific Revolution and the Foundations of Modern Science*. Westport: Greenwood, 2005.

Arbellot, Guy. Autour des routes de poste: les premières cartes routières de la France XVIIe–XIXe siècles. Paris: Bibliothèque Nationale, 1992.

Arbellot, Guy. "La grande mutation des routes de France au XVIIIe siècle," *Annales*, 28 (1973): 765–90.

Ariès, Philippe and Georges Duby, ed. *A History of Private Life*. Vol. 3: *Passions of the Renaissance*. Cambridge and London: The Belknap Press, 1989.

Ashworth, William B. "Natural History and the Emblematic World View." *Reappraisals of the Scientific Revolution*. Ed. David C. Lindberg and Robert S. Westman. Cambridge: Cambridge University Press, 1990, 303–32.

Aubrey, John. *Brief Lives*. Ed. Richard Barber. Woodbridge: The Boydell Press, 1982.

Avramescu, Cãtãlin. *An Intellectual History of Cannibalism*. Translated by Alistair Ian Blyth. Princeton and Oxford. Princeton University Press, 2009.

Babcock, Barbara, ed. *The Reversible World: Symbolic Inversion in Art and Society*. Ithaca: Cornell University Press, 1978.

Bacon, Francis. *The Collected Works of Francis Bacon*. Ed. James Spedding, Robert L. Ellis, and Douglas D. Heath. London: Longmans, 1878.

Bahlman, D. W. R. *The Moral Revolution of 1688*. New Haven: Yale University Press, 1957.

Bainton, Ronald. *Women of the Reformation: In Germany and Italy*. Philadelphia: Fortress Press, 1971.

Baker, Michael and Peter Hanns Reill, ed. *What's Left of Enlightenment?* Stanford: Stanford University Press, 2001.

Balibar, Renée et Dominque Laporte. *Le Français National: Politique et Pratiques de la Langue Nationale sous la Revolution Français*. Paris: Hachette Littérature, 1974.

Barber, John, ed. *John Aubrey Brief Lives*. Woodbridge: Boydell, 1982.

Baridon, Michel. "Les concepts de nature humaine et de perfectibilité dans l'histoiregraphie des Lumières de Fontenelle à Condorcet," *L'Histoire au dix-huitième siècle*. Actes du Colloque d'Aix-en-Provence. Aix-en-Provence: Edisud, 1980.

Barker-Benfield, G. J. *The Culture of Sensibility: Sex and Society in Eighteenth-Century Britain*. Chicago and London: The University of Chicago Press, 1992.

Barnett, S. J. *The Enlightenment and Religion: The Myths of Modernity*. Manchester and New York: Manchester University Press, 2003.

Barstow, Anne L. *Witchcraze: A New History of the European Witch Hunts*. San Francisco: Pandora, 1994.

Battis, Emery. *Saints and Sectarians: Anne Hutchinson and the Antinomian Controversy in the Massachusetts Bay Colony*. Chapel Hill: University of North Carolina Press, 1962.

Benavides, Gustavo. "Magic, Religion, Materiality," *History Reflections/Réflexions Historiques* 23 (1997): 301–30.

Benedict, Barbara M. *Curiosity: A Cultural History of Early Modern Inquiry*. Chicago and London: The University of Chicago Press, 2001.

Bennett, Jim. "Instruments and Illustrations in Eighteenth Century Astronomy." *Science and the Visual Image in the Enlightenment*. Ed. William R. Shea. Science History Publications/USA, 2000, 137–54.

Bennett, Jim. "The Mechanics' Philosophy and the Mechanical Philosophy," *History of Science* 24 (1986): 1–28.

Bennett, Jim and Scott Mandelbrote, ed. *The Garden, the Ark, the Tower, the Temple: Biblical Metaphors of Knowledge in Early Modern Europe*. Oxford: The Museum of the History of Science in association with the Bodleian Library, 1998.

Benz, Ernst. *Les sources mystique de la philosophie romantique*. Paris: Vrin, 1987 [1968].

Benz, Ernst. "Die Reinkarnationslehre in Dictung und Philosophie der deutschen Klassik und Romantik," *Zeitschrift für Religions-und Gesitesgeschichte* 9 (1957): 150–75.

Berger, Peter, ed. *The Desecularization of the World: Resurgent Religion and World Politics*. Grand Rapids, MI: William B. Eerdmans Publishing Company, 1999.

Birken, Lawrence. *Hitler as a Philosopher: Remnants of the Enlightenment in National Socialism*. Westport: Greenwood Press, 1995.

Blair, Ann. "Mosaic Physics and the Search for a Pious Natural Philosophy in the Late Renaissance," *Isis* 91 (2000): 32–58.

Blair, Ann. "Natural Philosophy." *The Cambridge History of Modern Science*. Vol. 3: *Early Modern Science*. Ed. Katherine Park and Lorraine Daston. Cambridge: Cambridge University Press, 2006, 365–406.

Blumenberg, Hans. *The Legitimacy of the Modern Age*. Translated by Robert M. Wallace. Cambridge: The MIT Press, 1983 [1966].

Bödeker, Hans Erich. "Menscheit, Humanität, Humanismus." *Geschichtliche Grundbegriffe: Historisches Lexikon zur politische-sozialen Sprache in Deutschland*. Ed. Otto Brunner, Werner Conze, und Reihart Koselleck, 8 vols. Stuttgart: E. Klett, 1972–97.

Boguet, Henri. *A Discourse on Witches*. London, 1602.

Bolzoni, Lina and Jeremy Parzen. *The Gallery of Memory: Literary and Iconographic Models in the Age of the Printing Press*. Toronto: University of Toronto Press, 2001.

Bono, James. *From Ficino to Descartes*, vol. 1 of *The Word of God and the Languages of Man: Interpreting Nature in Early Modern Science and Medicine*. Madison: the University of Wisconsin Press, 1995.

Borchardt, Frank L. "The Magus as Renaissance Man," *The Sixteenth Century Journal* 21 (Spring 1990): 57–76.

Bordo, Susan. *The Flight to Objectivity: Essays on Cartesianism and Culture*. Albany, NY: State University of New York Press, 1987.

Borst, Arno. *Der Turmbau von Babel: Geschichte der Meinungen über Ursprung und Vielfalt der Sprachen und Völker*, 6 vols. Stuttgart: Anton Hiersemann, 1957–63.

Bourdieu, Pierre. "Structures and the Habitus." *Outline of a Theory of Practice*. Translated by Richard Nice. Cambridge: Cambridge University Press, 1977, 72–95.

Bouwsma, William J. "Anxiety and the Formation of Early Modern Culture." *A Usable Past: Essays in European Cultural History*. Berkeley: University Of California Press, 1990, 157–89.

Bouwsma, William J. "Changing Assumptions in Later Renaissance Culture." *A Usable Past: Essays in European Cultural History*. 74–96.

Bouwsma, William J. *John Calvin: A Sixteenth Century Portrait*. New York and Oxford: Oxford University Press, 1988.

Bouwsma, William J. "The Secularization of Society in the Seventeenth Century." *A Usable Past: Essays in European Cultural History*, 112–28.

Bouwsma, William J. "The Two Faces of Humanism: Stoicism and Augustinianism in Renaissance Thought." *A Usable Past: Essays in European Cultural History*, 19–73.

Bouwsma, William J. *A Usable Past: Essays in European Cultural History*. Berkeley and Oxford; University of California Press, 1990.

Boyer, Paul and Stephen Nissenbaum. *Salem Possessed: The Social Origins of Witchcraft*. Cambridge: Harvard University Press, 1974.

Brann, Noel L. *Trithemius and Magical Theology*. Albany, NY: State University of New York Press, 1999.

Brauner, Sigrid. "Luther on Witchcraft: A True Reformer?" *The Politics of Gender in Early Modern Europe*. Ed. Jean R. Brink, Allison P. Coudert, Maryanne C. Horowitz. Kirksville: Sixteenth Century Journal Publishers, 1989, 33–34.

Brauner, Sigrid. *Fearless Wives and Frightened Shrews: The Construction of the Witch in Early Modern Germany*. Amherst: University of Massachusetts Press, 1995.

Bridgman, Annette. "Introduction." *Education in the 18th Century*. Ed. J. D. Browning. New York and London: Garland Publishing, 1979, 1–9.

Briggs, Robin. "The Science of Sin: Jacques de Sainte-Beuve and his *cas de conscience*." *Religious Change in Europe, 1660–1914*. Oxford: Oxford University Press, 1997, 23–40.,"

Brinsley, John. *A Looking-Glasse for Good Women*. London, 1645.

Broeckx, Corneille. "Interrogatoires du docteur J. B. van Helmontsur le magnétisme animal, publiés pour la première fois," *Annales de l'Academie d'archeologie de Belgique*, 13 (1856), 306–50.

Broedel, Hans Peter. *The Malleus Maleficarum and the Construction of Witchcraft. Theology and Popular Belief*. Manchester: Manchester University Press, 2003.

Brooke, John H. *Science and Religion: Some Historical Perspectives*. Cambridge: Cambridge University Press, 1991.

Brooke, John H. and Ian Maclean, ed. *Heterodoxy in Early Modern Science and Religion*. Oxford: Oxford University Press, 2005.

Brooke, John H., Margaret J. Osler, and Jitse M. van der Meer, ed. *Science in Theistic Contexts: Cognitive Dimensions* (*Osiris* 2.16). Chicago: Chicago University Press, 2001.

Brown, Peter. *Augustine of Hippo. A Biography*. Berkeley: University of California Press, 1969.

Brown, Peter. "Sorcery, Demons, and the Rise of Christianity from Late Antiquity into the Middle Ages." *Witchcraft Confessions and Accusations*. Ed. Mary Douglas. London: Tavistock Publications, 1970, 17–45.

Browne, Thomas. *Pseudodoxia Epidemica*. Ed. Robin Robbins, 2 vols. Oxford: Clarendon Press, 1981.

Burke, Peter. *Fortunes of the Courtier: The European Reception of Castiglione's Cortegiano*. Cambridge: Polity Press, 1995.

Burke, Peter, Brian Harrison, and Paul Slack, ed. *Civil Histories: Essays presented to Sir Keith Thomas*. Oxford: Oxford University Press, 2000.

Burke, Peter and Roy Porter, ed. *The Social History of Language*. Cambridge: Cambridge University Press, 1987.

Burke, Peter and Roy Porter, ed. *Language, Self, and Society: A Social History of Language*. Cambridge: Polity Press, 1991.

Burnet, Thomas. *Theory of the Earth*. London, 1697.

Butterfield, Herbert. *The Origins of Modern Science, 1300–1800*. 2nd ed. London: Bell, 1957.

Calvin, John. *Institutes of the Christian Religion*. 2 vols. Translated by John T. McNeil and Ford Lewis Battles. Philadelphia: The Westminister Press, 1960.

Cameron, Euan. *Enchanted Europe. Superstition, Reason, & Religion, 1250–1750*. Oxford: Oxford University Press, 2010.

Camporeale, Salvatore. "Renaissance Humanism and the Origins of Humanist Theology." *Humanity and Divinity in the Renaissance and Reformation: Essays in Honor of Charles Trinkhaus*. Ed. John W. O'Malley, Thomas M. Izbicki, and Gerald Christianson. Leiden: Brill, 1993, 101–124.

Casaubon, Meric. *A Treatise Concerning Enthusiasm*. London, 1656.

Cassirer, Ernst. *The Philosophy of the Enlightenment*. Princeton: Princeton University Press, 1979 [1932].

Castle, Terry. "The Culture of Travesty: Sexuality and Masquerade in Eighteenth-Century England." *British Literature, 1640–1789: A Critical Reader*. Ed. Robert Demaria, Jr. Oxford: Blackwell Publishing, 1999.

Castle, Terry. *The Female Thermometer: Eighteenth-Century Culture and the Invention of the Uncanny*. New York and Oxford: Oxford University Press, 1995.

Céard, Jean. *La Nature et les prodiges: L'Insolite au XVIe siècle en France*. Genève: Librairie Droz, 1977.

Certeau, Michel de. *La Fable mystique*. Paris: Gallimard, 1982; *The Mystic Fable*. Translated by Michael Smith. Chicago: University of Chicago Press, 1993.

Chatellier, Louis. *The Europe of the Devout: The Catholic Reformation and the Formation of a New Society*. Cambridge: Cambridge University Press, 1989 [1987].

Chesterfield, Philip Dormer Stanhope, Earl of. *Lord Chesterfield's Letters*. Ed. David Roberts. Oxford: Oxford University Press.

Clark, Stuart. "The 'Gendering' of Witchcraft in French Demonology: Misogyny or Polarity?" *French History* 5 (1991), 426–37.

Clark, Stuart. "The Scientific Status of Demonology." *Occult & Scientific Mentalities in the Renaissance*. Ed. Brian Vickers. Cambridge: Cambridge University Press, 1984, 351–74.

Clark, Stuart. *Thinking with Demons: The Idea of Witchcraft in Early Modern Europe*. Oxford: Clarendon Press, 1997.

Clark, Stuart. "Witchcraft and Magic in Early Modern Culture." *Witchcraft and Magic in Europe. The Period of the Witch Trials*. Ed. Bengt Ankarloo and Stuart Clark. Philadelphia: University of Pennsylvania Press, 2002, 96–169.

Clark, William, Jan Golinski and Simon Schaffer, ed. *The Sciences in Enlightenment Europe*. Chicago: University of Chicago Press, 1999.

Clarke, Samuel. *A Demonstration of the Being and Attributes of God*. London, 1704.

Clarkson, Leslie. *Death, Disease and Famine in Pre-Industrial England*. New York: St Martin's Press, 1975.

Clericuzio, Antonio. "From van Helmont to Boyle: A Study of the Transmission of Helmontian Chemical and Medical Theories in Seventeenth-Century England," *British Journal for the History of Science* 26 (1993): 303–43.

Clericuzio, Antonio. "A Redefinition of Boyle's Chemistry and Corpuscular Philosophy," *Annals of Science* 47 (1990): 561–89.

Clericuzio, Antonio. "Robert Boyle and the English Helmontians." *Alchemy Revisited. Proceedings of the International Conference on the History of Alchemy at the University of Groningen*. Ed. Z. R. W. M. von Martels. Leiden: Brill, 1990.

Clulee, Nicholas H. *John Dee's Natural Philosophy: Between Science and Religion*. London: Routledge, 1988.

Cockaye, Emily. *Hubbub: Filth, Noise & Stench in England, 1600–1770*. New Haven: Yale University Press, 2007.

Cody, Lisa Forman, "The Doctor's in Labour; or a New Whim Wham from Guilford," *Gender and History* 4 (1992): 175–96.

Cohen, H. Floris. *The Scientific Revolution: A Historiographical Inquiry*. Chicago and London: University of Chicago Press, 1994.

Cohen, I. B. *Revolution in Science*. Cambridge: Belknap Press & Harvard University Press, 1985.

Cohn, Norman. *The Pursuit of the Millennium: Revolutionary Millenarianism and Mystical Anarchists of the Middle Ages*. London: Secker & Warburg, 1957.

Cohen, Sherrill. "Asylums for Women in Counter-Reformation Italy." *Women in Reformation and Counter-Reformation Europe: Private and Public Worlds*. Ed. Sherrin Marshall. Bloomington and Indianapolis: Indiana University Press, 1989, 166–188.

Contreras, Jaime and Gustav Henningsen, "Forty-four Thousand Cases of the Spanish Inquisition (1540–1700): Analysis of a Historical Data Bank." *The Inquisition in Early Modern Europe*. Studies on Sources and Methods. Ed. Gustav Henningsen and J. Tedschi. Illinois: Northern Illinois University Press, 1986.

Conway, Anne. *The Principles of the Most Ancient and Modern Philosophy* (1690). Ed. Allison P. Coudert and Taylor Corse. Cambridge Texts in the History of Philosophy. Cambridge: Cambridge University Press, 1996.

Cook, Harold J. "Medicine." *The Cambridge History of Science*. Vol. 3: Early Modern Science. Ed Katharine Park and Lorraine Daston. Cambridge: Cambridge University Press, 2006, 408–34.

Cooper, Thomas. *The Mystery of Witchcraft*. London, 1617.

Copenhaver, Brian P. "Hermes Trismegistus, Proclus and the Question of a Theory of Magic in the Renaissance." *Hermeticism and the Renaissance: Intellectual History and the Occult in Early Modern Europe*. Ed. Ingrid Merket and Allen G. Debus. Washington: Folger Library, 1989, 79–110.

Copenhaver, Brian P. *Hermetica: The Greek Corpus Hermeticum and the Latin Asclepius*. Translated by Brian P. Copenhaver. Cambridge: Cambridge University Press, 1992.

Copenhaver, Brian P. "Natural Magic, Hermetism, and Occultism." *Reappraisals of the Scientific Revolution*. Ed. David C. Lindberg and Robert S. Westman. Cambridge: Cambridge University Press 1990, 261–301.

Copenhaver, Brian P. "The Occultist Tradition and its Critics." *The Cambridge History of Seventeenth Century Philosophy*. Ed. Daniel Garber and Michael Ayers. 2 vols. Cambridge: Cambridge University Press, 1998, 2: 454–512

Cornelius, Paul. *Languages in Seventeenth and Early Eighteenth Century Imaginary Voyages*. Genèvre: Librairie Droz, 1965.

Correll, Barbara. *The End of Conduct: Grobianus and the Renaissance Text of the Subject*. Ithaca and London: Cornell University Press, 1996.

Costa, Milagros Oretga. "Spanish Women in the Reformation. *Women in Reformation and Counter-Reformation Europe: Public and Private Worlds*. Ed. Sherrill Marshall. Bloomington: Indiana University Press, 1989, 89–119.

Coudert, Allison P. "Educating Girls in Early Modern Europe and America." *Childhood in the Middle Ages and the Renaissance: the Results of a Paradigm Shift in the History of Mentality*. Ed. Albrecht Classen. Berlin & New York: Walter de Gruyter, 2005, 389–413.

Coudert, Allison P. "Faith and Religion." *A Cultural History of Childhood and Family in the Age of the Enlightenment*. Ed. Elizabeth Foyster and James Marten. *A Cultural History of Childhood and Family*. 6 vols. Ed. Elizabeth Elizabeth Foyster and James Marten. Oxford and New York: Berg, 2010, vol. 4, 147–164.

Coudert, Allison P. *The Impact of the Kabbalah in the Seventeenth Century: The Life and Thought of Francis Mercury van Helmont, 1614–1698*. Leiden: Brill, 1998.

Coudert, Allison P. "The Kabbalah, Science, and the Enlightenment: The Doctrines of *Gilgul* and *Tikkun* as Factors in the Anthropological Revolution of the Eighteenth Century," *Aufklärung und Esoterik: Rezeption—Integration—Konfrontation*. Ed. Monika Neugebauer-Wölk. Tübingen: Max Niemeyer, 2008, 299–316.

Coudert, Allison P. "Seventeenth Century Natural Philosophy and Esotericism at the Court of Sulzbach." *Mélanges d'Histoire des Religions réunis en l'honneur de M. Antoine Faivre par ses éleves, collègues et amis*. Ed. Joscelyn Godwin and Wouter J. Hanegraaff. Louvain: Peeters, 2000, 27–46.

Coudert, Allison P. "Sewers, Cesspools, and Privies: Waste as Reality and Metaphor in Pre-modern European Cities." *Urban Space in the Middle Ages and the Early Modern Age*. Ed. Albrecht Classen and Marilyn Sandidge. Berlin and New York: Walter de Gruyter, 2009, 713–33.

Coudert, Allison P., Sarah Hutton, Richard H. Popkin, and Gordon M. Weiner, *Judaeo-Christian Intellectual Culture in the Seventeenth Century: A Celebration of the Library of Narcissus Marsh*. Dordrecht: Kluwer, 1999.

Coudert, Allison P. and Jeffrey S. Shoulson, ed. *Hebraica Veritas? Christian Hebraists, Jews, and the Study of Judaism in Early Modern Europe*. Philadelphia: University of Pennsylvania Press, 2004.

Coupe, William A. *The German Illustrated Broadsheet in the Seventeenth Century*. 2 vols. Historical and Iconographical Studies. Bibliotheca Bibliographica Aureliana 17. Baden-Baden: Verlag Librarie Heitz GMBH, 1966.

Couturat, Louis. *La Logique de Leibniz*. Hildesheim: Olms, 1961 [1901].

Cox, John C. *The Parish Registers of England*. London: Methuen, 1910.

Cressy, David. *Bonfires and Bells: The National Memory and the Protestant Calendar in Elizabethan and Stuart England*. London: Weidenfeld and Nicolson, 1989.

Cross, Claire. "Great Reasoners in Scripture: The Activities of Women Lollards, 1380–1530. *Medieval Women*. Ed. Derek Baker. Oxford: Oxford University Press, 1978.

Cunningham, Andrew. "Getting the Game Right. Some Plain Words on the Identity and Invention of Science," *Studies in the History and Philosophy of Science* 19 (1988): 365–89.

Cunningham, Andrew and Perry William. "De-centering the 'Big Picture': The Origins of Modern Science and the Modern Origins of Science," *British Journal for the History of Science* 26 (1993): 407–32.

Dagens, Jean. *Bérulle et les origins de la restauration catholique*. Paris: Bruges, 1952.

D'Alembert, Jean Le Rond. *Preliminary Discourse to the Encycopedia of Diderot*. Translated by Richard N. Schwab. Chicago: University of Chicago Press, 1995.

Daneau, Lambert. *The Wonderful Workmanship of the World*. Translated by Thomas Twyne. Facsimile edition. Whitefish: Kessinger Publishing, 2010.

Darnton, Robert. *Mesmerism and the End of the Enlightenment in France*. New York: Schocken Books, 1968.

Darnton, Robert. "Readers Respond to Rousseau: The Fabrication of Romantic Sensitivity." *The Great Cat Massacre and other Episodes in French Cultural History*. New York: Vintage Book, 1985, 215–63.

Daston, Lorraine and Katherine Park. "Hermaphrodites in Renaissance France," *Critical Matrix* 1 (1985): 1–19.

Davis, Natalie Z. *Society and Culture in Early Modern France. Eight Essays*. Stanford: Stanford University Press, 1975.

Dear, Peter. "Miracles, Experiments and the Ordinary Course of Nature," *Isis* 81 (1990): 663–83.

Dear, Peter. "Nullius in verba," " 'Totius in verba': Rhetoric and Authority in the Early Royal Society," *Isis* 76 (1985): 145–61.

Dear, Peter. *Revolutionizing the Sciences: European Knowledge and Its Ambitions, 1500–1700*. 2nd ed. Princeton University Press, 2009.

Debrock, G. "Aristotle, Wittgenstein, *alias* Isaac Newton between Fact and Substance." *Newton's Science and Philosophical Legacy*. Ed. P. B. Schurer and G. Debrock. Dordrect: Kluwer, 1988, 355–77.

Debus, Allen G. *The Chemical Philosophy: Paracelsian Science and Medicine in the Sixteenth and Seventeenth Centuries*. 2 vols. New York: Science History Publications, 1977.

Debus, Allen G. *Chemistry, Alchemy and the New Philosophy, 1550–1700: Studies in the History of Science and Medecine*. London: Variorum Reprints, 1987.

Debus, Allen G. *The English Paracelsians*. New York: F. Watts, 1966.

Debus, Allen G. *The French Paracelsians*. Cambridge: Cambridge University Press, 1992.

Debus, Allen G. *Science and Education in the Seventeenth Century: The Webster-Ward Debate*. London: Macdonald, 1970.

Deleuze, Gilles. *The Fold: Leibniz and the Baroque*. Minneapolis: University of Minnesota Press, 1993.

Demos, John P. *Entertaining Satan: Witchcraft and The Culture of Early New England*. New York and Oxford: Oxford University Press, 1982.

Demos, John P. "Underlying Themes in the Witchcraft of Seventeenth Century New England," *The American Historical Review* 75 (1970), 1311–26.

Demott, Benjamin. "Comenius and the Real Character in England," *PMLA* 70 (1955): 1068–81

Descartes, René. *The Philosophical Writings of Descartes*. Translated by John Cottingham, Robert Stoothoff, and Dugdale Murdoch. 3 vols. Cambridge: Cambridge University Press, 1985.

Dobbs, Betty Jo Teeter. *The Foundations of Newton's Alchemy: Or, "The Hunting of the Greene Lyon."* Cambridge: Cambridge University Press, 1975.

Dobbs, Betty Jo Teeter. *The Janus Faces of Genius: The Role of Alchemy in Newton's Thought*. Cambridge: Cambridge University Press, 1991.

Dobbs, Betty Jo Teeter. "Newton as Final Cause and First Mover." *Rethinking the Scientific Revolution*. Ed. Margaret J. Osler. Cambridge: Cambridge University Press, 2000, 25–40.

Donne, John. *The Complete English Poems*. Ed. Albert James Smith. London: Penguin, 1986.

Dooley, Brendan. *Morandi's Last Prophecy and the End of Renaissance Politics*. Princeton and Oxford: Princeton University Press, 2002.

Douglas, Mary. "The Abominations of Leviticus." *Purity and Danger: An Analysis of Concepts of Pollution and Taboo*. London and New York: Routledge, 2002 [1966], 41–57.

Douglas, Mary. "Deciphering a Meal." *Implicit Meanings: Selected Essays in Anthropology*. 2nd ed. London and New York: Routledge, 1999 [1975], 231–51.

Dresen-Coenders, Lène, ed. *Saints and She-devils. Images of Women in the 15th and 16th centuries*. Translated by C. M. H. Sion and R. M. J. van der Wilden. London: The Rubicon Press, 1987.

Dunbar, Robin. *The Trouble with Science*. Cambridge: Harvard University Press, 1996.

Eamon, William. *Science and the Secrets of Nature: Books of Secrets in Medieval and Early Modern Culture*. Princeton: Princeton University Press, 1994.

Eamon, William. "Technology as Magic in the Late Middle Ages and the Renaissance," *Janus* 70 (1983): 171–212.

Eby, Frederick. *Early Protestant Educators: The e Educational Writings of Martin Luther, John Calvin, and Other Leaders of Protestant Thought*. New York: AMS Press, 1971 [1931].

Ecco, Umberto. *The Search for the Perfect Language*. Translated by James Fentress. Oxford: Blackwell, 1995.

Ehnmaark, Erland. "Religion and Magic—Frazer, Söderblom, and Hägerström," *Ethnos* 21 (1952): 1–10.

Eisenstein, Elizabeth L. *The Printing Press as an Agent of Change. Communications and Cultural Transformations in Early Modern Europe*. Cambridge and New York: Cambridge University Press, 1979.

Elias, Norbert. *The Civilizing Process*. Translated by Edmund Jephcott. Oxford: Blackwell Publishing, 2000 [rev. ed.].

Elliott, John H. *The Old World and The New, 1492–1650*. Cambridge: Cambridge University Press, 1970.

Erasmus, Desiderius. "On Good Manners for Boys/ De Civilitate Morum puerlium" (1530).Translated and annotated by Brian McGregor. *Collected Works of Erasmus*. Toronto: University of Toronto Press, 1985.

Euben, Roxanne L. *Islamic Fundamentalism and the Limits of Modern Rationalism: A Work of Comparative Political Theory*. Princeton: Princeton University Press, 1999.

Evans, R. J. W. *Rudolph II and His World: A Study in Intellectual History, 1576–1612*. Oxford: Oxford University Press, 1973.

Evans, R. J. W. "Learned Societies in Germany in the Seventeenth Century," *European Studies Review* 7 (1977): 129–51.

Fara, Patricia. *Newton: The Making of a Genius*. New York: Columbia University Press, 2002.

Feingold, Mordechai, ed. *History of Universities*, Vol. 20/2. Oxford: Oxford University Press, 2005.

Feingold, Mordechai, ed. *Jesuit Science and the Republic of Letters*. Cambridge: MIT Press, 2003.

Feldhay, Rivka. "Religion." *The Cambridge History of Modern Science*. Vol. 3: *Early Modern Science*. Cambridge; Cambridge University Press, 2006, 727–55.

Ficino, Marsilio. *Opera Omnia*. Basle, 1576.

Filmer, Robert. *Patriarcha and Other Political Writings*. Ed. Peter Laslett. Oxford: Blackwell, 1949.

Findlen, Paula. *Possessing Nature: Museums, Collecting, and Scientific Culture in Early Modern Italy*. Berkeley: University of California Press, 1994.

Findlen, Paula, ed. *Athanasius Kircher: The Last Man Who Knew Everything*. London and New York: Routledge, 2004.

Finke, Roger and Rodney Stark, *The Churching of America 1776–1990: Winners and Losers in our Religious Economy*. New Brunswick: Rutgers University Press, 2002.

Flandrin, Jean-Louis. *Families in former times: kinship, household and sexuality*. Cambridge: Cambridge University Press, 1979 [1976].

Foucault, Michel. *Discipline and Punish: The Birth of the Prison*. New York: Viking, 1975.

Foucault, Michel. *The Order of Things: An Archaeology of the Human Sciences*. New York: Vintage Books, 1973 [1966].

Foucault, Michel. "What is Critique?" *What is Enlightenment: Eighteenth Century Answers and Twentieth Century Questions*. Ed. James Schmidt. Berkeley: University of California Press 1996, 382–98.

Fox, George. *Fox's Book of Miracles*. Ed. Henry J. Cadbury. Cambridge: Cambridge University Press, 1948.

Foyster, Elizabeth A. *Manhood in Early Modern England*. London: Longman Publishing Group, 1999.

Frazer, James G. *The Golden Bough* (abridged edition). New York: Macmillan, 1963 [1922].

Friedman, Jerome. *The Battle of the Frogs and Fairford's Flies: Miracles and the Pulp Press During the English Revolution*. London: UCL Press, 1993.

Frost, Alan. "The Pacific Ocean: The Eighteenth-Century's 'New World,' " *Studies in Voltaire and the Eighteenth Century* 142 (1976): 279–322."

Gardner, John. *The Life and Times of Chaucer*. New York: Knopf, 1977.

Garin, Eugenio. "La 'Dignitas hominis' e la letteratura patristica," *Rinascita* 1 (1938): 102–46.

Gatti, Hilary. *Giordano Bruno and Renaissance Science*. Ithaca: Cornell University Press, 1999.

Gatti, Hilary. *The Renaissance Drama of Knowledge: Giordano Bruno in England*. London and New York: Routledge, 1989.

Gaukroger, Stephen. *The Emergence of a Scientific Culture: Science and the Shaping of Modernity, 1210–1685*. Oxford: Clarendon Press, 2006.

Gaukroger, Stephen. *Francis Bacon and the Transformation of Early-Modern Philosophy*. Cambridge: Cambridge University Press, 2001.

Gay, Peter. "The Enlightenment as Medicine and as Cure." *The Age of Enlightenment: Studies Presented to Theodore Besterman*. Ed. William H. Barbour. Edinburgh: St. Andrews University Publications, 1967, 375–86.

Gay, Peter. *The Enlightenment: An Interpretation*. Vol. 1: *The Rise of Modern Paganism*. London: Norton, 1966.

Geertz, Clifford. "Towards an Interpretive Theory of Culture." *The Interpretation of Cultures*. New York: Basic Books, 1973.

Geertz, Hilda. "An Anthropology of Religion and Magic," *Journal of Interdisciplinary History* 6 (1975): 71–89.

Giammatti, A. Bartlett. "Proteus Unbound: Some Versions of the Sea God in the Renaissance." *The Disciplines of Criticism: Essays in Literary Theory, Interpretation, and History*. Ed. Peter Demetz, Thomas Greene, and Lowry Nelson, Jr. New Haven: Yale University Press, 1968, 437–75.

Gierl, Martin. *Pietismus und Aufklarung: theologische Polemik und die Kommunikationsreform der Wissenschaft am Ende des 17. Jahrhunderts*. Göttingen: Vandenhoeck & Ruprecht, 1995.

Gillespie, Neal C. "Natural History, Natural Theology, and The Social Order: John Ray and the 'Newtonian Ideology,'" *Journal of the History of Biology* 20 (1987): 1–49.

Ginzburg, Carlo. *Ecstasies: Deciphering the Witches' Sabbath*. Translated by Raymond Rosenthal. New York: Pantheon, 1991.

Ginzburg, Carlo. "High and Low: The Theme of Forbidden Knowledge in the Sixteenth and Seventeenth Centuries," *Past and Present* 73 (1976): 28–41.

Girard, Allain. "Le triomphe de la cuisinère bourgeoisie: livres culinaires, cuisine et société en France, XVIIe et XVIIIe siècles," *Revue d'histoire moderne et contemporaine* (1977): 497–512.

Glanvill, Joseph. *Lux Orientalis*. . . . London, 1662.

Glanvill, Joseph. *Saducismus triumphatus* . . . London, 1681.

Gliozzi, Giuliano. *Adam et il Nuovo Mondo: la Nascita dell'Anthropologia come Ideologia Coloniale*. Firenze: La Nuova Italia, 1977.

Glucklich, Ariel. *The End of Magic*. New York and Oxford: Oxford University Press, 1997.

Goldberg, Jonathan. "Fatherly Authority: The Politics of Stuart Family Images." *Rewriting the Renaissance: The Discourse of Sexual Difference in Early Modern Europe*. Ed. Margaret W. Ferguson, Maureen Quilligan & Nancy J. Vickers. Chicago: University of Chicago Press, 1986, 3–32.

Goldammer, Kurt. "Paracelsische Eschatologie: zum Verständnis der Anthroplogie und Kosmologie Hohenheims," *Nova Acta Paracelsica* 5 (1948): 45–85.

Goldammer, Kurt. "Paracelsische Eschatologie II, *Nova Acta Paracelsica* 6 (1952): 68–102.

Gombrich, Ernst H. "Icones Symbolicae: The Visual Image in Neo-Platonic Thought," *Journal of the Warburg and Courtauld Institutes* 11 (1948), 163–92; expanded version in idem, *Symbolic Images*. London: Phaidon, 1972.

Goodman, Lenn E., ed. *Neoplatonism in Jewish Thought*. Albany: State University of New York Press, 1992.

Goodman, Lenn E., ed. *Jewish and Islamic Philosophy: Crosspollination in the Classic Age*. Piscataway: Rutgers University Press, 1999.

Gorceix, Bernard. *La Bible des Rose-Croix*. Paris: Presses Universitaires, 1970.

Gouge, William. *Domesticall Duties*. London, 162).

Grafton, Anthony. *Defenders of the Text: The Traditions of Scholarship in an Age of Science, 1450–1800*. Cambridge: Harvard University Press, 1994.

Grafton, Anthony. "The Importance of Being Printed," *Journal of Interdisciplinary History* 11 (1980/81): 265–86.

Grafton, Anthony. *Joseph Scaliger; A Study in the History of Classical Scholarship*. Oxford: Oxford University Press, 1983.

Grafton, Anthony. "Momigliano's Method and the Warburg Institute: Studies in his Middle Period." *Momigliano and Antiquarianism: Foundations of the Modern Cultural Sciences*. Ed. P. N. Miller. Toronto: University of Toronto Press, 2007, 97–126.

Grafton, Anthony (with April Shelford and Nancy Siraisi), *New Worlds, Ancient Texts: The Power of Tradition and the Shock of Discovery*. Cambridge: The Belknap Press of Harvard University Press, 1992.

Gray, John. *Enlightenment's Wake: Politics and Culture at the Close of the Modern Age*. London: Routledge, 1995.

Greenblatt, Stephen. *Marvelous Possessions: The Wonder of the New World*. Oxford: Clarendon Press, 1991.

Gregory, Tullio. "Pierre Charron's 'Scandalous Book.' " *Atheism from the Refomation to the Enlightenment*. Ed. Michael Hunter and David Wotten. Oxford: Clarendon Press, 1992.

Greven, Philip. *The Protestant Temperament. Patterns of Child-Rearing, Religious Experience and the Self in Early America*. New York: Alfred A. Knopf, 1977.

Grimm, Heinrich. "Die deutschen 'Teufelbucher' des 16 Jahrhuunderts: ihre rolle im Buchwesen und ihre Bedeutung," *Archiv für Geschichte des Buchwesens* 16 (1959): 513–70.

Habermas, Jürgen. *The New Conservatism*. Translated by Shierry W. Nicholson and Richard Wolin. Cambridge: MIT Press, 1991.

Habermas, Jürgen. *Structural Transformation of the Public Sphere: Inquiry into a Category of Bougeois Society*. New York: Polity Press, 1992 [1962].

Hall, David D., ed. *The Antinomian Controversy, 1636–38: A Documentary History*. Durham: Duke University Press, 1990.

Haller, William and Malleville Haller. "The Puritan Art of Love," *Huntington Library Quarterly* (1942): 235–72.

Halleux, Robert. "Theory and Experiment in the Early Writings of Johan Baptista Van Helmont." *Theory and Experiment*. Ed. Diderik Batens and Jean Paul Van Bendegem.. Dordrecht: Reidel, 1988, 93–101.

Hannaway, Owen. *Chemists and the Word*. Baltimore: Johns Hopkins University Press, 1975.

Hanegraaff, Wouter J. "Beyond the Yates Paradigm: The Study of Western Esotericism Between Counterculture and New Complexity," *Aries: Journal for the Study of Western Esotericism*, NS 1 (2001): 5–37.

Hanegraaff, Wouter J. "On the Construction of 'Esoteric Traditions.' " *Western Esotericism and the Science of Religion*. Ed. Antoine Faivre &Wouter J. Hanegraaff. Leuven: Peeters, 1998.

Hanegraaff, Wouter J. and Joyce Pijnenburg, ed. *Hermes in the Academy: Ten Years' Study of Western Esotericism at the University of Amsterdam*. Amsterdam: Amsterdam University Press, 2009.

Harkness, Deborah E. *John Dee's Conversations with Angels: Cabala, Alchemy, and the End of Nature*. Cambridge: Cambridge University Press, 1999.

Harris, Steven J. "Transposing the Merton Thesis: Apostolic Spirituality and the Establishment of the Jesuit Scientific Tradition," *Science in Context* 3 (1989): 29–65.

Harris, Victor. *All Coherence Gone. A Study of the Seventeenth Century Controversy over Disorder and Decay in the Universe*. London, 1966 [1949].

Harrison, Peter. *The Bible, Protestantism, and the Rise of Natural Science*. Cambridge: Cambridge University Press, 1998.

Harrison, Peter. "Curiosity, Forbidden Knowledge, and the Reformation of Natural Philosophy in Early Modern England," *Isis*, 92 (2001): 265–90.

Harrison, Peter. *The Fall of Man and the Foundations of Science*. Cambridge: Cambridge University Press 2007.

Harrison, Peter. "Original Sin and the Problem of Knowledge in Early Modern Europe," *Journal of the History of Ideas* 63 (2002): 239–60.

Hayden, Hiram. *The Counter*-Renaissance. New York: Charles Scribner's sons, 1950.

Heer, Friedrich. *Die Dritte Kraft: Der europäische Humanismus zwischen den Fronten des konfessionellen Zeitalters*. Frankfurt am Main: S. Fischer Verlag, 1960.

Heilbron, John. "Domesticating Science in the Eighteenth Century." *Science and the Visual Image in the Enlightenment*. Ed. William R. Shea. Science History Publications/USA, 2000, 1–24.

Hein, Hilde. "The Endurance of the Mechanism-Vitalism Controversy," *Journal of the History of Biology* 5 (1972): 159–88.

Hellyer, Marcus. *Catholic Physics: Jesuit Natural Philosophy in Early Modern Germany*. Notre Dame: University of Notre Dame Press, 2005.

Helmont, Francis Mercury van. *The Alphabet of Nature*. A translation with an introduction of Francis Mercury van Helmont's *Alphabeti vere naturalis Hebraici brevissima Delineatio*. ... [Sulzbach, 1667]. Ed. and translated by Allison P. Coudert and Tayor Corse. Leiden and Boston: Brill, 2007.

Helmont, J. B. *Deliramenta Catarrhi* ... Translated by Walter Charleton. London, 1650.

Helmont, J. B. *Oriatrike or, Physick Refined*. ... Translated by Walter Charleton. London, 1662.

Helmont, J. B. *Ortus Medicinae*. ... Ed. F. M. van Helmont. Amsterdam, 1648.

Helmont, J. B. *A Ternary of Paradoxes*. ... Translated by Walter Charleton. London, 1650.

Hendrix, Scott H. "Martin Luther and Albrecht of Mainz," *Lutherjahrbuch* 49 (1982): 96–114.

Henningsen, Gustav. *The Witches' Advocate: Basque Witchcraft and the Spanish Inquisition*. Reno: University of Nevada Press, 1980.

Henry, John. "Occult Qualities and the Experimental Philosophy: Active Principles in Pre-Newtonian Matter Theory," *History of Science* 24 (1986): 335–81.

Henry, John. *The Scientific Revolution and the Origins of Modern Science*. 3rd. ed. Houndmills: Palgrave Macmillan, 2008 [1997, 2002].

Heseler, Baldasar. *Andreas Versalius' First Public Anatomy at Bologna 1540: An Eyewitness Report*. Ed. Ruben Eriksson. Uppsala: Almquist & Wiksells, 1959.

Heyd, Micheal. *"Be Sober and Reasonable." The Critique of Enthusiasm in the Seventeenth and Early Eighteenth Centuries*. Leiden: E. J. Brill, 1995.

Hill, John. A *Review of the Works of the Royal Society, containing animadversions on such papers as deserve particular observation*. London, 1751.

Hinman, Robert B. "The Apotheosis of Faust: Poetry and New Philosophy in the Seventeenth Century." *Metaphysical Poetry*. Ed. D. J. Palmer & M. Bradbury. Stratford–upon–Avon Studies no. 11. London: Edward Arnold, 1970, 149–79.

Hoak, Dale. "Witch Hunting and Women in the Art of the Renaissance," *History Today* 31 (1981): 22–26.

Hobbes, Thomas. *Leviathan*. Ed. Michael Oakeshott. New York: Collier, 1962.

Hollinger, David A. "The Enlightenment and the Genealogy of Cultural Conflict in the United States." *What is Left of the Enlightenment?* Ed. Keith Michael Baker and Peter Hanns Reill. Stanford: Stanford University Press, 2001, 7–18.

Hoppen, Theodore K. "The Nature of the Early Royal Society, pt. 1," *The British Journal for the History of Science* 9 (1976), 1–24.

Horkheimer, Max and Theodor W. Adorno. *Dialetic of Enlightenment*. Translated by John Cumming. New York: Continuum, 2002 [1944].

Horowitz, Maryanne C. "Aristotle and Women," *Journal of the History of Biology* 9 (1979):182–213.

Houghton, Walter B. "The English Virtuoso in the Seventeenth Century, Pt. I and Pt. II," *Journal of the History of Ideas* 3.1 & 2 (1942): 51–73, 190–219.

Hull, Isabel V. *Sexuality, State, and Civil Society in Germany, 1700–1815*. Ithaca and London: Cornell University Press, 1996.

Hults, Linda C. "Baldung and the Reformation." *Hans Baldung Grien: Prints and Drawings*. Ed. James H. Marrow and Alan Shestack. New Haven: Yale University Art Gallery, 1981, 38–59.

Hults, Linda C. "Baldung and the Witches of Freiburg: The Evidence of Images," *Journal of Interdisciplinary History* 18 (19878): 249–76.

Hunter, Michael and David Wootton, ed. *Atheism in Early Modern Europe*. Oxford: Oxford University Press, 1992.

Hunter, Michael. *Establishing the New Science: The Experience of the Early Royal Society*. Woodbridge: Boydell Press, 1989.

Hunter, Michael. "Science and Heterodoxy: An Early Modern Problem Reconsidered." *Reappraisals of the Scientific Revolution.* Ed. David. C. Lindberg and Robert S. Westman. Cambridge: Cambridge University Press, 1990, 437–60.

Hunter, Michael. *Science and Society in Restoration England.* Cambridge: Cambridge University Press, 1981.

Hunter, Michael, ed. *The Occult Laboratory: Magic, Science and Second Sight in Late 17th-Century Scotland: A new edition of Robert Kirk's The Secret Commonwealth and other texts, with an introductory essay*. Woodbridge: The Boydell Press, 2001.

Hutchinson, James C. "The Housebook Master and the Folly of the Wise Man," *Art Bulletin* 48 (1966): 73–78.

Hutchinson, Keith. "What Happened to Occult Qualities in the Scientific Revolution?" *Isis* 73 (1982): 233–53.

Hutton, Ronald. *The Rise and Fall of Merry England: The Ritual Year, 1400–1700.* Oxford: Oxford University Press, 1994.

Idel, Moshe. *New Perspectives on the Kabbalah.* New Haven: Yale University Press, 1988.

Iliffe, Robert. " 'Is He Like Other Men?' The Meaning of the *Principia Mathematica* and the Author as Idol." *Culture and Society in the Stuart Restoration.* Ed. Gerald Maclean. Cambridge: Cambridge University Press, 1995, 159–76.

Impey, Oliver and Arthur MacGregor, ed. *The Origin of Museums: The Cabinet of Curiosities in Sixteenth and Seventeenth Century Europe.* Oxford: Clarendon Press, 1985.

Israel, Jonathan I. *Radical Enlightenment: Philosophy and the Making of Modernity, 1650–1750.* Oxford: Oxford University Press, 2002.

Jacob, Margaret. "The Truth of Newton's Science and the Truth of Science's History: Heroic Science at Its Eighteenth-Century Formulation." *Rethinking the Scientific Revolution.* Ed. Margaret J. Osler. Cambridge: Cambridge University Press, 2000, pp. 315–32.

Jacob, Margaret and Dorothée Sturkenboom. "A Woman's Scientific Society in the West: The Late Eighteenth Century Assimilation of Science," *Isis* 94 (2003): 3217–52.

Jardine, Lisa and Alan Stewart. *Hostage to Fortune: The Troubled Life of Francis Bacon* London: Gollancz, 1998.

Jardine, Nicholas, James A. Secord, and Emma C. Spary, ed. *The Cultures of Natural History*. Cambridge: Cambridge University Press, 1995.

Johns, Adrian. *The Nature of the Book: Print and Knowledge in the Making.* Chicago: Chicago University Press, 1998.

Johnson L. W. and M. L. Wolbarsht. "Mercury Poisoning: A Probable Cause of Isaac Newton's Physical and Mental Ills," *Notes and Records of the Royal Society* 34 (1970): 1–9.

Johnston, Trevor. "The Reformation and Popular Culture." *The Reformation World*. Ed. Andrew Petegree. London: Routledge, 2000, 545–60.

Jolly, Karen. "Medieval Magic: Definitions, Beliefs, Practices." *Witchcraft and Magic in Europe. The Middle Ages*. Ed. Bengt Ankarloo and Stuart Clark. Philadelphia: University of Pennsylvania Press, 2002.

Jones, Richard F. *Ancients and Moderns: A Study of the Rise of the Scientific Movement in Seventeenth Century England*. New York: Dover, 1982 [1961].

Josselin, Ralph. *The Diary of Ralph Josselin, 1616–1683*. Ed. Alan MacFarlane. Oxford: Oxford University Press, 1976.

Jung, Carl Gustav. *Alchemical Studies, The Collected Works of C. G. Jung*. Ed. Herbert Read, Michael Fordam, Gerhard Adler, and William McGuire. Translated by R. F. C. Hull. Vol. 13. London: Routledge, 1967.

Jung, Carl Gustav. *Mysterium Coniunctionis: An Inquiry into the Separation and Synthesis of Psychic Opposites in Alchemy. The Collected Works of C. G. Jung*. Vol. 14.

Kahn, Coppélia. *Man's Estate: Masculine Identity in Shakespeare*. Berkeley: University of California Press, 1981.

Kaplan, Barbara Beigun. "Greatrakes, the Stoker: The Interpretation of his Contemporaries," *Isis* 73 (1982): 178–83.

Karant-Nunn, Susan C. "Continuity and Change: Some Effects of the Reformation on the Women of Zwickau," *Sixteenth Century Journal* 12 (1982).

Karant-Nunn, Susan C. "Reformation Society, Women and the Family." *The Reformation World*. Ed. Andrew Pettegree. London and New York: Routledge, 2000, 433–60.

Karlsen, Carol. *The Devil in the Shape of a Woman: Women and Witchcraft in Colonial New England*. New York: W. W. Norton, 1987.

Keller, Evelyn Fox. *Reflections on Gender and Science*. New Haven: Yale University Press, 1985.

Kelley, Donald H. *Foundations of Modern Historical Scholarship: Language, Law and History in the French Renaissance*. New York and London: Columbia University Press, 1970.

Kelso, Ruth. *Doctrine for the Lady of the Renaissance*. Urbana: University of Illinois Press, 1956.

Kieckhefer, Richard. "The Specific Rationality of Medieval Magic," *The American Historical Review* 99 (1994): 813–36.

Kilgour, Maggie, *From Communion to Cannibalism: An Anatomy of Metaphors of Incorporation*. Princeton: Princeton University Press, 1990.

King, Winston L. *Introduction to Religion*. New York: Harper and Brothers, 1954.

Kirsch, Irving. "Demonology and Science during the Scientific Revolution," *Journal of the History of the Behavioral Sciences* 8 (1980): 359–68

Kittredge, George Lyman. *Witchcraft in Old and New England*. Cambridge: Harvard University Press, 1928.

Klaits, Joseph. *Servants of Satan: The Age of the Witch Hunt*. Bloomington: University of Indiana Press, 1985.

Klaaren, Eugene M. *Religious Origins of Modern Science: Belief in Creation in Seventeenth-Century Thought*. Grand Rapids: Eerdmans, 1977.

Kleinberg, Aviad M. " *De agone christiano*: The Preacher and His Audience," *Journal of Theological Studies* 38 (1987): 16–33.

Koch, Robert A. *Hans Baldung Grien; Eve, the Serpent and Death*. Chicago: University of Chicago Press, 1974.

Knowlson, James. *Universal Language Schemes in England and France, 1600–1700*. Toronto: University of Toronto Press, 1975.

Koyré, Alexandre. *From the Closed World to the Infinite Universe*. Baltimore: Johns Hopkins University Press, 1957.

Kramer Heinrich and James Sprenger, *Malleus Maleficarum*. Ed. and translated by P. G. Maxwell-Stuart. Manchester and New York; Manchester University Press, 2007.

Kristeller, Paul O. *Renaissance Thought and Its Sources*. Ed. Michael Mooney. New York: Columbia University Press, 1979, Kubrin, David. "Newton and the Cyclical Cosmos: Providence and the Mechanical Philosophy," *Journal of the History of Ideas* 28 (1967): 325–46.

Kunzle, David. *The Early Comic Strip*. Berkeley: University of California Press, 1973.

Langmuir, Gavin. *Towards a Definition of Antisemitism*. Berkeley: University of California Press, 1990.

Laporte, Dominique. *History of Shit*. Translated by Nadia Benabid and Rudolphe el-Khoury with and introduction by Rudolphe el-Khoury. Cambridge: MIT Press, 2000 [1978].

Laqueur, Thomas. *Making Sex: Body and Gender from the Greeks to Freud*. Cambridge: Harvard University Press, 1990.

Laqueur, Thomas. *Solitary Sex: A Cultural History of Masturbation*. New York: Zone Books, 2003.

Larner, Christina. *Enemies of God: The Witch-hunt in Scotland*. London: Chatto & Windus, 1981.

Latour, Bruno. *The Pasteurization of France*. Translated by Alan Sheridan and John Law. Cambridge and London: Harvard University Press, 1988.

Leff, Gordon. *Medieval Thought: St. Augustine to Ockham*. London: The Merlin Press, 1959.

Lehrich, Christopher I. *The Language of Demons and Angels: Cornelius Agrippa's Occult Philosophy*. Leiden: Brill, 2003.

Leibniz, Gottfried Wilhelm. *Tagebuch, Geschichtliche Aufsätze und Gedichte Gottfried Wilhelm Leibniz Gesammeltewerk*. Ed. G. H. Pertz ([1847] Hildesheim: Olms, 1966).

Lemaius, Levinus. *An Herbal for the Bible*. Translated by Thomas Newton. London, 1587.

Leonard, Amy E. "Female Religious Orders." *A Companion to the Reformation World*. Ed. R. Po-chia Hsia. Oxford: Blackwell Publishers, 2004, 237–54.

Levack, Brian P. *The Witch-Hunt in Early Modern Europe*. London and New York: Longman, 1987.

Levin, Harry. *The Myth of the Gold Age in the Renaissance*. Bloomington: Indiana University Press, 1969.

Levine, Joseph M. *The Battle of the Books: History and Literature in the Augustan Age*. Ithaca: Cornell University Press, 1991.

Lindberg, Carter. *The European Reformations*. Oxford: Blackwell Publishing, 1996.

Lindberg, David C. and Ronald L. Numbers, ed. *God & Nature: Historical Essays on the Encounter between Christianity and Science*. Berkeley: University of California Press, 1986.

Lindberg, David C. *When Science and Christianity Meet*. Chicago: Chicago University Press, 2003.

Lindemann, Mary. *Health & Healing in Eighteenth Century Germany*. Baltimore: Johns Hopkins University Press, 1996.

Lowthorpe, John, ed. *The Philosophical Transactions and Collections to the End of the Year 1700, Abridg'd and Dispos'd under General Heads*. 2nd ed., London, 1716.

Luther, Martin. *Luther's Works*. Ed. Jaroslave Pelikan. Philadelphia: Concordia Publishing Company, 1968.

Lüthy, Christoph. "The Confessionalization of Physics: Heresies, Facts and the Travails of the Republic of Letters." *Heterodoxy in Early Modern Science and Religion*. Ed. John Brooke and Ian Maclean. Oxford: Oxford University Press, 2005, 81–114.

McCalla, Arthur. "Palingenesie Philosophique to Palingenesis Sociale: From a Scientific Ideology to a Historical Ideology," *Journal of the History of Ideas* 55 (1994): 421–39.

McClure, Ruth. *Coram's Children: The London Foundling Hospital in the Eighteenth Century*. New Haven: Yale University Press, 1981.

Mcfarlane, Alan. *Witchcraft in Tudor and Stuart England: A Regional and Comparative Study*. London: Routledge & Kegan Paul, 1970.

McGuire, J. E. "Neoplatonism and Active Principles: Newton and the *Corpus Hermeticium*." Robert S. Westman & J. E. McGuire, *Hermeticism and the Scientific Revolution*. Berkeley and Los Angeles: University of California Press, 1977, 95–142.

McKeon, Michael. *The Origins of the English Novel, 1600–1740*. Baltimore and London: The Johns Hopkins Press, 2002 [1987].

McLaughlin, "Equality of Souls, Inequality of Sexes: Women in Medieval Theology." *Religion and Sexism: Images of Woman in the Jewish and Christian Traditions*. Ed. Rosemary R. Ruether. New York: Simon and Schuster, 1974, 213–66.

McManners, John. *Death and the Enlightenment: Changing Attitudes to Death among Christians and Unbelievers in Eighteenth-century France*. Oxford: Clarendon Press, 1981.

Maier, Michael *Atalanta Fugiens: hoc est, Emblemata nova de secretis naturae chymica* . . . Oppenheim, 1618. *Michael Maier's Atalanta Fugiens: An Edition of the Emblems, Fugues and Epigrams*. Translated and edited by Joscelyn Godwin. Magnum Opus Hermetic Sourceworks #22. Grand Rapids: Phanes Press, 1989.

Markley, Robert. *Fallen Language: Crises of Representation in Newtonian England, 1660–1740*. Ithaca: Cornell University Press, 1993.

Mather, Cotton. *Diary of Cotton Mather, 1681–1708*. Massachusetts Historical Society Collections, vol. 7. Boston: Published by the Society, 1911.

Maxwell-Stuart, P. G. *The Occult in Early Modern Europe: A Documentary History*. Basingstoke: Macmillan, 1999.

Mebane, John S. *Renaissance Magic and the Return of the Golden Age: The Occult Tradition and Marlowe, Jonson, and Shakespeare*. Lincoln: University of Nebraska Press, 1989.

Medick, Hans. "Village Spinning Bees: Sexual Culture and Free Time among Rural Youth in Early Modern Germany." *Interest and Emotion: Essays on the Study of Family and Kinship*. Ed. Hans Medick and David W. Sabean. Cambridge: Cambridge University Press, 1984.

Mellinkoff, Ruth. "Riding Backwards: Theme of Humiliation and Symbol of Evil," *Viator* 4 (1973): 153–86.

Merchant, Carolyn. *The Death of Nature*. San Francisco: Harper & Row, 1983 [1980].

Merchant, Carolyn. *Earthcare: Women and the Environment*. New York: Routledge, 1995.

Merton, Robert K. "Puritanism, Pietism and Science," *Sociological Review* 28 (1936): 1–30.

Merton, Robert K. "Science, Technology and Society in Seventeenth-Century England," *Osiris* 4 (1938): 360–632 [rpt New York: Howard Fertig/Harpter Torchbooks, 1970].

Meyer, R. W. *Leibniz and the Seventeenth-Century Revolution.* Translated by J. P. Stern. Chicago: Henry Regnery Company, 1952.

Mévergnies, Paul Nève de. *Jean-Baptiste van Helmont: Philosophe par le Feu.* Paris: Librarie Droz, 1935.

Middleton, W. E. K. "What did Charles II Call the Fellows of the Royal Society?" *Notes and Records* 32 (1977);

Midelfort, H. C. Eric. "Heartland of the Witchcraze: Central and Northern Europe," *History Today* 21 (1981): 27–31.

Midelfort, H. C. *Witch-hunting in South Western Germany, 1562–1684: The Social and Intellectual Foundations.* Stanford: Stanford University Press, 1972.

Miller, Peter. "The 'Antiquarianization' of Biblical Scholarship and the London Polyglot Bible (1653–57)," *The Journal of the History of Ideas* 62 (July 2001): 463–82.

Miller, Peter. "Making the Paris Polyglot Bible: Humanism and Orientalism in the early Seventeenth Century." *Die europäische Gelehrtenrepublikim Zeitalter des Konfessionalismus/The European Republic of Lettersin the Age of Confessionalism.* Ed. H. Jaumann. Wiesbaden: Harrassowitz, 2001, 59–86.

Miller, Peter. *Momigliano and Antiquarianism: Foundations of the Modern Cultural Sciences.* Toronto: University of Toronto Press, 2007.

Momigliano, Arnoldo. *The Classical Foundations of Modern Historiography.* Berkeley: University of California Press, 1990.

Monter, E. William."Inflation and Witchcraft: The Case of Jean Bodin." *Action and Conviction in Early Modern Europe.* Ed. Theodore K. Rabb and J. J. Siegel. Princeton: Princeton University Press, 1969.

Monter, E. William. "Pedestal and Stake: Courtly Love and Witchcraft." *Becoming Visible: Women in European History.* Ed. Renate Bridenthal and Claudia Koonz. Boston: Houghton Mifflin, 1977.

Monter, E. William. *Witchcraft in France and Switzerland: The Borderlands During the Reformation.* Ithaca: Cornell University Press, 1976.

Monter, E. William. "Witch Trials in Continental Europe." *Witchcraft and Magic in Europe. The Period of the Witch Trials.* Ed. Bengt Ankarloo and Stuart Clark. Philadelphia: University of Pennsylvania Press, 2002, 3–52.

Monter, E. William. "Women in Calvinist Geneva (1550–1800)," *Signs* 6 (1980): 189–209.

Montrose, Louis. "A Midsummer Night's Dream and the Shaping Fantasies of Elizabethan Culture: Gender, Power, Form." *Rewriting the Renaissance: The Discourse of Sexual Difference in Early Modern Europe.* Ed. Margaret W. Ferguson, Maureen Quilligan & Nancy J. Vickers. Chicago: University of Chicago Press, 1986.

Montrose, Louis. "Shaping Fantasies and Figurations of Gender and Power in Elizabethan Culture," *Representations* 2 (Spring 1983): 61–94.

Moore, Cornelia Niekus. *The Maiden's Mirror: Reading Material of German Girls in the Sixteenth and Seventeenth Centuries*. Wolfenbüttler Forschungen. Wiesbaden: Otto Harrassowitz, 1977.

Moran, Bruce T. *The Alchemical World of the German Court: Occult Philosophy and Chemical Medicine in the Circle of Moritz of Hessen (1572–1632)*. *Sudhoff's Archiv*, Beiheft 29. Stuttgart: Franz Steiner Verlag 1991.

Moran, Bruce T. "Privilege, Communication, and Chemiatry: The Hermetick–Alchemical Circle of Moritz of Hessen-Kassel," *Ambix* 32 (1985): 110–27.

Moran, Bruce T., ed. *Patronage and Institutions: Science, Technology, and Medicine at the European Courts, 1500–1750*. Woodbridge: Boydell Press, 1991.

Moran, Bruce T., ed. *Distilling Knowledge: Alchemy, Chemistry, and the Scientific Revolution*. Cambridge: Harvard University Press, 2006.

More, Henry. *Conjectura Cabbalistica*. London, 1653.

Morgan, Edmund S. *The Puritan Family: Religion & Domestic Relations in 17th Century New England*. New York: Harper & Row, 1966.

Moxey, K. *Peasants, Warriors, and Wives. Popular Imagery in the Reformation*. Chicago: Chicago University Press, 1989.

Mulligan, Lotte. "Anglicanism, Latitudinarianism and Science in Seventeenth Century England," *Annals of Science*, 30 (1973): 213–19.

Mulligan, Lotte. "Civil War Politics, Religion and the Royal Society," *Past and Present* 59 (1973): 92–116.

Mulligan, Lotte. "Science, Religion and Politics, a Rejoinder," *Past and Present* 66 (1975): 39.

Mulsow, Martin. "Vernünftige Metempsychosis: Über Monadenlehre, Esoterik und geheime Aufklärungsgesellschaften im 18. Jahrhundert." *Aufklärung und Esoterik*. Ed. Monika Neubebauer-Wölk.Hamburg: Felix Meiner Verlag, 1999, 211–73.

Nadelhaft, Jerome. "The Englishwoman's Sexual Civil War: Feminist Attitudes Towards Men, Women, and Marriage, 1650–1740," *The Journal of the History of Ideas* 43 (1982): 555–79.

Needham, Joseph. *A History of Embryology*, 2d ed., rev. New York: Abelard-Schuman, 1959.

Neiman, Susan. *Evil in Modern Thought*. Princeton: Princeton University Press, 2002.

Neugebauer-Wölk, Monika, ed. *Aufklärung und Esoterik*. Hamburg: Felix Meiner Verlag, 1999.

Newman, William R. "Alchemy, Domination, and Gender," *A House Built on Sand: Exposing Postmodernist Myths about Science*. Ed. NorettaKoertge. New York: Oxford University Press, 1998.

Newman, William R. "Experimental Corpuscular Theory in Aristotelian Alchemy." *Late Medieval and Early Modern Corpuscular Matter Theories*. Ed. Christoph Lüthy, John E. Murdoch, and William R. Newman. Leiden: Brill, 2001.

Newman, William R. *Promethean Ambitions: Alchemy and the Quest to Perfect Nature*. Chicago and London: The University of Chicago Press, 2004.

Newman, William R. "Technology and Alchemical Debate in the Late Middle Ages," *Isis* 80 (1989): 423–45.

Newman, William R. and Lawrence M. Principe. "Alchemy vs. Chemistry: The Etymological Origins of a Historiographic Mistake," *Early Science and Medicine* 3 (1998): 32–65

Newman, William R. and Lawrence M. Principe. *Alchemy Tried in the Fire: Starkey, Boyle, and the Fate of Helmontian Chymistry*. Chicago: University of Chicago Press, 2002.

Nicolson, Marjorie H. "Virtuoso." *Dictionary of the History of Ideas*. Ed. Philip P. Wiener. 5 vols. New York: Charles Scribner and Sons, 1973–74, 4: 486–90.

Nicolson, Marjorie H., ed. *Conway Letters* (1930). Revised and edited by Sarah Hutton. Oxford: Clarendon Press, 1992.

Oberman, Heiko A. *Luther: Man between God and the Devil.* Translated by Eileen Walliser-Schwarzbat. New Haven: Yale University Press, 1989.

Obeyesekere, Gananath. *Cannibal Talk: The Man-Eating Myth and Human Sacrifice in the South Seas*. Berkeley: The University of California Press, 2005.

O'Malley, John. "The Society of Jesus," *A Companion to the Reformation*. Ed. R. Po-chia Hsia. Oxford: Blackwell Publishing, 2004, 223–36.

Ong, Walter J. *Rhetoric, Romance, and Technology: Studies in the Interaction of Expression and Culture*. Ithaca and London: Cornell University Press, 1971.

Orgel, Stephen. "Prospero's Wife." *Rewriting the Renaissance: The Discourse of Sexual Difference in Early Modern Europe*. Ed. Margaret W. Ferguson, Maureen Quilligan & Nancy J. Vickers. Chicago: University of Chicago Press, 1986.

Osborn, Max. *Die Teufelliteratur des XVI Jahrhunders*. Hildesheim: Georg Olms, 1965.

Osler, Margaret J., ed. *Rethinking The Scientific Revolution*. Cambridge: Cambridge University Press, 2000.

Ozment, Steven. *When Fathers Ruled: Family Life in Reformation Europe*. Cambridge: Harvard University Press, 1983.

Pagden, Anthony. *The Fall of Natural Man: The American Indian and the Origins of Comparative Ethnology*. Cambridge: Cambridge University Press, 1982.

Pagel, Walter. "The Debt of Science and Medicine to a Devout Belief in God," *Journal of the Transactions of the Victoria Institute* 74 (1942): 100–116.

Pagel, Walter. *Joan Baptiste van Helmont: Reformer of Science and Medicine*. Cambridge: Cambridge University Press, 1982.

Pagel, Walter. *Paracelsus: An Introduction to Philosophical Medecine in the Era of the Renaissance*. 2nd rev. ed. Basel and New York: Karger, 1982 [1958].

Pagel, Walter. "Paracelsus and the Neoplatonic and Gnostic Tradition." *Religion and Neoplatonism in Renaissance Medicine*. Ed. Marianne Winder. London: Variorum Reprints, 1985.

Pagel, Walter. "The Religious and Philosophical Aspects of van Helmont's Science and Medicine," *Bulletin of the History of Medicine*, supp. 2 (1944).

Pagel, Walter. "Religious Motives in the Medical Biology of the XVIIth. Century," *Bulletin of the History of Medicine* 3 (1935), 97–128.

Pagel, Walter. *William Harvey's Biological Ideas*. Basel: Karger, 1958.

Pagels, Elaine. *Adam, Eve, & the Serpent*. New York: Random House, 1988.

Pagels, Elaine. *The Gnostic Gospels*. New York: Vintage Books, 1979

Park, Katherine. "Medicine and Magic: The Healing Arts." *Gender and Society in Renaissance Italy*. Ed. Judith C. Brown and Robert C. Davis. London: Longman, 1998.

Park, Katherine and Lorraine Daston, ed. *The Cambridge History of Modern Science*. Vol. 3: *Early Modern Science*. Cambridge: Cambridge University Press, 2006.

Parry, Graham. *The Golden Age Restored: The Culture of the Stuart Court, 1603–42*. Manchester: Manchester University Press, 1985.

Partington, J.R. "Joan Baptista van Helmont," *Annales of Science* 1 (1936): 368.

Patrides, C.A. "The Salvation of Satan," *The Journal of the History of Ideas* 28 (1967): 467–78.

Paul, Charles B. *Science and Immortality: The Éloges of the Paris Academy of Sciences, 1699–1791*. Berkeley: University of California Press, 1980.

Paulson, Ronald. *Hogarth's Harlot: Sacred Parody in Enlightenment England*. Baltimore and London: Johns Hopkins University Press, 2003.

Pears, Iain. *An Instance of the Finger Post*. New York: Riverhead Books, 1998.

Pelt, R. J. van. *Tempel van de Wereld: de Kosmische Symboliek von de Tempel van Salomo*. Utrecht: HES, 1984.

Pelt, R. J. van. "Through the Temple of Solomon to the Temple of Heaven: Bacon's *New Atlantis*," *Boletin de Arte* 2 (1981): 45–52.

Pesic, Peter. "Westling with Proteus: Francis Bacon and the 'Torture' of Nature," *Isis* 90 (1999): 81–94.

Peter, J. P. "Disease and the Sick at the End of the Eighteenth Century." *Biology of Man in History*. Ed. Robert Forster and Orest Ranum. Translated by Elborg Forster and Patricia M. Ranum. Baltimore and London, 1975.

Petrarch, Francesco. *Letters to Familiar Authors*. Ed. Mario Cosenza. Chicago; Chicago University Press, 1910.

Pettersson, Olof. "Magic—Religion: Some Marginal Notes to an Old Problem," *Ethnos* 22 (1957): 109–19.

Petty, William. *The Petty Papers*, Ed. The Marquis of Landsdowne. 2 vols. London: Constable, 1927.

Peters, E. "Libertas Inquirendi and the Vitum Curiositatis in Medieval Thought." *La notion de liberté au Moyen Age: Islam, Bysance, Occident*. Ed. George Makdishi, Dominique Sourdel, and Janine Sourdel-Thomine. Paris: Les Belles Lettres, 1985, 89–98.

Pingree, David, ed. *Picatrix: The Latin Version of the Ghâyat Al-Hakîm*. London, Warburg Institute, 1986.

Polyani, Karl. *The Great Transformation: The Political and Economic Origins of Our Time*. New York: Farrar & Rinehart, 1944.

Popkin, Richard H. *The History of Scepticism: From Savonarola to Bayle*. Oxford: Oxford University Press, 2003.

Popkin, Richard H. "The Religious Background of Seventeenth Century Philosophy," *Journal of the History of Philosophy* 25 (1987): 35–50

Popkin, Richard H. *The Third Force in Seventeenth Century Thought*. Leiden and New York: E. J. Brill, 1992.

Porter, Roy. *Flesh in the Age of Reason: The Modern Foundation of Body and Soul*. New York and London: W. W Norton & Company, 2003.

Principe, Lawrence M. *The Aspiring Adept: Robert Boyle and his Alchemical Quest*. Princeton: Princeton University Press, 1998.

Quaife, G. R. *Wanton Wenches and Wayward Wives: Peasants and Illicit Sex in Early Seventeenth Century England*. New Brunswick: Rutgers University Press, 1979.

Rattansi, P. "Paracelsus and the Puritan Revolution," *Ambix* 11 (1963): 24–32.

Real, Hermann Josef. "Tradition and Progress in Bacon's Advancement of Learning. *Scientia potestas est*—Knowledge is Power: Francis Bacon to Michel Foucault," *Anglia* 119 (2001): 1–19.

Reill, Peter H. *The German Enlightenment and the Rise of Historicism*. Berkeley: University of California Press, 1975.

Reinburg, Virginia. "Hearing Lay People's Prayer." *Culture and Identity in Early Modern Europe, 1500–1800: Essays in Honor of Natalie Zemon Davis*. Ed. Barbara Diefendorf and Carla Hesse. Ann Arbor: University of Michigan Press, 1993, 19–39

Reinhard, Wolfgang. "Gegenreformation als Modernisierung? Prolegomena zu einer Theorie des konfessionellen Zeitalters," *Archivfür Reformationsgeschichte* 68 (1977): 226–52.

Reinhard, Wolfgang. "Reformation, Counter-Reformation and the Early Modern State: a Reassessment," *Catholic Historical Review* 75 (1989): 383–404.

Ricken, Ulrich. *Linguistics, Anthropology and Philosophy in the French Enlightenment: Language, Theory and Ideology*. Translated by Robert E. Norton. London: Routledge: 1994 [1984].

Ricoeur, Paul. *The Symbolism of Evil*. New York: Harper and Row, 1967.

Roberts, Penny. "Agencies Human and Divine: Fire in French Cities, 1520–1720." *Fear in Early Modern Society*. Ed. William G. Naphy and Penny Roberts. Manchester: Manchester University Press, 1997.

Roos, Keith L. *The Devil in Sixteenth-Century German Literature: The Teufelsbücher*. Bern: Peter Lang, 1972.

Roper, Lyndal. *The Holy Household: Women and Morals in Reformation Augsburg*. Oxford Studies in Social History. Oxford: Clarendon Press, 1989.

Roper, Lyndal. "Luther: Sex, Marriage and Motherhood," *History Today* (December 1983).

Roper, Lyndal. *Oedipus & the Devil: Witchcraft, Sexuality, and Religion in Early Modern Europe*. New York: Routledge, 1994.

Rosenau, Helen. *Vision of the Temple: The Image of the Temple in Judaism and Christianity*. London: Oresko Books, 1979.

Ross, G. MacDonald, *Leibniz*. Oxford: Oxford University Press, 1984.

Ross, G. "Occultism and Philosophy in the Seventeenth Century." *Philosophy, Its History and Historiography*. Ed. A. J. Holland. Dordrecht: D. Reidel, 1985, 95–115.

Rossi, Paolo. *The Dark Abyss of Time: This History of the Earth & The History of Nations from Hooke to Vico*. Translated by Lydia G. Cochrane. Chicago: The University of Chicago Press, 1984.

Rossi, Paolo. *Francis Bacon: From Magic to Science*. Translated by Sacha Rabinovitch. Chicago: Chicago University Press, 1968.

Rossi, Paolo. *Logic and the Art of Memory: The Quest for a Universal Language*. Translated with an introduction by Stephen Clucas. London: Athlone Press, 2000.

Russell, Bertrand and Paul Edwards. *Why I am not a Christian; Together with Other Essays on Religion and Related Subjects*. London: Allen &Unwin, 1967.

Russell, H. Diane. *Eva/Ave: Women in Renaissance and Baroque Prints*. New York: Feminist Press at the City of New York, 1990.

Ryan, Michael T. "Assimilating New Worlds in the Sixteenth and Seventeenth Centuries," *Comparative Studies in Society and History* 23 (1981): 519–38.

Sagan, Carl. *The Demon Haunted World: Science as a Candle in the Dark*. London: Headline, 1996.

Said, Edward. *Culture and Imperialism*. London: Chatto & Windus, 1993.

Saint-Evremont, Charles de. *The Works of Monsieur St. Evremond, with the Life of the Author by Des Maizeaux*. Edited and translated by Pierre Des Maizeux. 2nd ed., London, 1728.

Saler, Michael. "Modernity and Enchantment: A Historiographic Review," *The America Historical Review* 111 (2006): 692–716.

Salmon, Vivian. "Language Planning in the Seventeenth Century: Its Context and Aims. *The Study of Language in 17th Century England*. Amsterdam: Benjamins, 1979.

Sarton, George. "Aristotle and Phyllis," *Isis* 14 (1930): 8–19.

Sarton, George. *Introduction to the History of Science*. 3 vols. Baltimore: Williams and Wilkins, 1927–47.

Scaffer, Simon. "Godly Men and Mechanical Philosophers: Souls and Spirits in Restoration Philosophy," *Science in Context* 1 (1978); 55–85.

Scaffer, Simon. "Occultism and Reason." *Philosophy, Its History and Historiography*. Ed. A. J. Holland. Dordrecht: D. Reidel, 1985, 117–43.

Schama, Simon. *Citizens: A Chronicle of the French Revolution*. New York: Alfred A. Knopf, 1989.

Schama, Simon. *The Embarrassment of Riches: An Interpretation of Dutch Culture in the Golden Age*. Berkeley: University of California Press, 1988.

Schiebinger, Londa. *The Mind Has No Sex? Women in the Origins of Modern Science*. Cambridge: Harvard University Press, 1989.

Schilling, Heinz. "Confessional Europe." *Handbook of European History 1400–1600*. Ed. Thomas A Brady, Jr., Heiko Oberman, and James D. Tracy. Leiden, Brill, 1995.

Schilling, Heinz. "Confessionalisation and the Rise of Religious and Cultural Frontiers in Early Modern Europe." *Frontiers of Faith: Religious Exchange and the Constitution of Religious Identities, 1400–1750*. Ed. Eszter Andor and Istávan G. Tóth. Budapest: Central European University and European Science Foundation, 2001, 21–36.

Schilling, Heinz. "Fruheneuzeitliche Formierung und Disziplinierung von Ehe, Familie und Erziehungim Spiegel calvinistischer Kirchenratsprotokolle." *Glaube und Eid: Treueformeln, Glaubensbekenntnisse und Sozialdisziplinierungzwischen Mittlelater und Neuzeit*. Ed. Paolo Prodi. Munich: Oldenbourg, 1993.

Schmidt, James, ed. *What is Enlightenment? Eighteenth-Century Answers and Twentieth-Century Questions*. Berkely, Los Angeles and London: University of California Press, 1996.

Schmitt, Charles B. *Aristotle and the Renaissance*. Cambridge: Cambridge University Press, 1983.

Schmitt, Charles B. *The Aristotelian Tradition and Renaissance Universities*. London: Variorum, 1984.

Schmitt, Charles B. *John Case and Aristotelianism in the Renaissance*. Kingston: McGill-Queen's University Press, 1983.

Schmitt, Charles B. "Reappraisals in Renaissance Science," *History of Science* 16 (1978).

Schmitt, Charles B. "Towards a Reassessment of Renaissance Aristotelianism," *History of Science* 11 (1973).

Schmitt, Charles, ed. *History of Universities*. Vol. 14. Oxford: Oxford University Press, 1998.

Schnucker, Robert V. "La position puritaine à l'egard de l'adulte," *Annales: Économies, Sociétés, Civilisations* 27 (1972): 1379–88.

Schuchard, Marsha Keith. *Restoring the Temple of Vision. Cabalistic Freemasonry and Stuart Culture*. Leiden: Brill, 2002.

Schücking, Levin L. *The Puritan Family: A Social Study from the Literary Sources*. Translated by Brian Battershaw. New York: Schocken Books, 1979 [1929].

Schuler, Robert. "Spiritual Alchemies of Seventeenth Century England," *The Journal of the History of Ideas* 41 (1980): 293–318.

Schutte, Anne J. " 'Such Monstrous Births': A Neglected Aspect of the Antinomian Controversty," *Renaissance Quarterly* 38 (Spring 1985): 85–106.

Scott, Joan W. "Gender: A Useful Category of Historical Analysis," *American Historical Review* 91 (December 1986): 1053–75.

Scott, Joan W. *Gender and the Politics of History*. New York: Columbia University Press, 1999 [1989].

Scribner, Robert W. "Incombustible Luther: The Image of the Reformer in Early Modern Europe," *Past & Present* 110 (1986), 38–68.

Scribner, Robert W. "The Reformation, Popular Magic, and the 'Disenchantment of the World,' " *Journal of Interdisciplinary History* 23 (1993): 475–94.

Secret, François. "Palingenesis, Alchemy and Metempsychosis in Renaissance Medicine," *Ambix* 26 (1979): 81–99.

Seigel, Jerrold E. *The Idea of the Self: Thought and Experience in Western Europe*. Cambridge: Cambridge University Press, 2005.

Sewell, Samuel. *Diary of Samuel Sewell, 1674-1729*. Boston: Massachusetts Historical Society, 1878–82.

Shadwell, Thomas. *The Virtuoso*. Ed. Marjorie Hope Nicolson and Stuart Rodes. Lincoln: University of Nebraska Press, 1966.

Shapin, Steven. *The Scientific Revolution*. Chicago and London: The University of Chicago Press, 1998 [1996].

Shapin, Steven. "Understanding the Merton Thesis," *Isis* 79 (1988): 594–605.

Shapin, Steven and Simon Schaffer. *Leviathan and the Air Pump*. Princeton: Princeton University Press, 1985.

Shapiro, Barbara. Shapiro, *John Wilkins*. Los Angeles: University of California Press, 1969.

Shapiro, Barbara. "Latitudinarianism and Science in Seventeenth-Century England," *The Intellectual Revolution of the Seventeenth Century*. London; Routledge & Kegan Paul, ed. Charles Webster, 1974, 286–316.

Shapiro, Barbara. *Probability and Certainty in Seventeenth-Century England: A Study of the Relationships between Natural Science, Religion, History, Law and Literature*. Princeton: Princeton University Press, 1983.

Sharpe, J. R. "The History of Crime in Late Medieval and Early Modern England: A Review of the Field," *Social History* 7 (1982): 197–203.

Shaw, Jane. *Miracles in Enlightenment England*. New Haven and London: Yale University Press, 2006, 97.

Sheehan, Jonathan. *The Enlightenment Bible: Translation, Scholarship, Culture*. Princeton: Princeton University Press, 2005.

Sheehan, Jonathan. "The Enlightenment, Religion, and the Enigma of Secularization: A Review Essay," *The American Historical Review* 108 (October 2003), 1061–80.

Shelton, Anthony Alan. "Cabinets of Transgression: Renaissance Collections and the Incorporation of the New World." *The Culture of Collecting*. Cambridge: Harvard University Press, 1994.

Shestack, Alan. "An Introduction to Hans Baldung." Ed. James H. Marrow and Alan Shestack. New Haven: Yale University Art Gallery, 1981.

Shorto, Russell. *Descartes' Bones: A Skeletal History of the Conflict between Faith and Reason*. New York; Doubleday, 2008.

Shuger, Debora K. *Habits of Thought in the English Renaissance: Religion, Politics, and the Dominant Culture*. Berkeley: University of California Press, 1990.

Shuger, Debora K. *The Renaissance Bible: Scholarship, Sacrifice, and Subjectivity* Berkley: University of California Press, 1998 [1994].

Simon, Richard. *A Critical History of the Old Testament*. London, 1682.

Slaughter, Margaret M. *Universal Language and Scientific Taxonomy in the Seventeenth Century*. Cambridge: Cambridge University Press, 1982.

Smith, Jonathan Z. "The Influence of Symbols on Social Change: A Place on Which to Stand." *Map is Not Territory. Studies in the History of Religions*. Chicago and London: University of Chicago Press, 1978, 129–46.

Smith, Moira. "The Flying Phallus and the Laughing Inquisitor: Penis Theft in the *Malleus Maleficarum*," *Journal of Folklore Research* 39:1 (2003): 85–121.

Smith, Pamela. *The Business of Alchemy*. Princeton: Princeton University Press, 1994.

Smith, Pamela. *The Body of the Artisan: Art and Experience in the Scientific Revolution*. Chicago: University of Chicago Press, 2004.

Sobel, Dava. *Longitude: The True Story of a Lone Genius Who Solved the Greatest Scientific Problem of His Time*. Penguin, 1996.

Soergel, Philip. "Miracles." *The Oxford Encyclopedia of the Reformation*. Vol. 3. Ed. Hans J. Hillerbrand. New York and Oxford: Oxford University Press, 1996.

Soergel, Philip. "Miracle, Magic, and Disenchantment in Early Modern Germany." *Envisioning Magic: A Princeton Seminar & Symposium*. Ed. Peter Schäfer and Hans G. Kippenber. Leiden: Brill, 1997, 215–34

Soergel, Philip. *Wondrous in His Saints: Propaganda for the Catholic Reformation in Bavaria*. Berkeley: University of California Press, 1993.

Spargo P. E. and C. A. Pounds. "Newton's 'Derangement of the Intellect': New Light on an Old Problem," *Notes and Records of the Royal Society* 34 (1970):11–32.

Spary, Emma C. *Utopia's Garden: French Natural History from Old Regime to Revolution*. Chicago: University of Chicago Press, 2000.

Spinoza, Baruch. *Tractatus Theological-Politicus*. Translated by Samuel Shirley with an introduction by Brad S. Gregory. Leiden: Brill, 1989.

Spiro, Melford E. "Religion: Problems of Definition and Explanation." *Anthropological Approaches to the Study of Religion*. Ed. Michael Banton. London: Tavistock Publications, 1966.

Sprat, Thomas, *The History of the Royal Society*. London, 1667.

Stallybrass, Peter. "The World Turned Upside Down: Inversion, Gender and the State." *The Matter of Difference: Materialist Feminine Criticism of Shakespeare*. Ed. V. Wayne. Ithaca: Cornell University Press, 1991, 201–17.

Stark, Rodney. *For the Glory of God: How Monotheism Led to the Reformations, Science, Witch-Hunts, and the End of Slavery*. Princeton: Princeton University Press, 2003.

Stark, Rodney and William Sims Bainbridge. *A Theory of Religion*. New Brunswick: Rutgers University Press, 1996 [1987].

Steiner, Arpad. "The Faust Legend and the Christian Tradition," *Proceedings of the Modern Language Association* 54 (1939), 391–404.

Stephen, James. "Philanthropy and Empire: Jonas Hanway and the Infant Poor of London," *Eighteenth-Century Studies* 12 (1979): 285–305

Stephens, Russell. "Witches who Steal Penises: Impotence and Illusion in the *Malleus Maleficarum*," *Journal of Medieval and Early Modern Studies* 28 (1998): 495–529.

Stephens, Walter. *Demon Lovers: Witchcraft, Sex, and the Crisis of Belief*. Chicago: University of Chicago Press, 2002.

Stewart, Larry. *The Rise of Public Science: Rhetoric, Technology, and Natural Philosophy in Newtonian Britain, 1660–1750*. Cambridge: Cambridge University Press, 1992.

Stollberg, Michael. "A Woman Down to Her Bones: The Anatomy of Sexual Difference in the Sixteenth and Early Seventeenth Centuries," *Isis* 94 (2003): 274–99.

Stolzenberg, Daniel. *Kircher's Hieroglyphs: Erudition and Occult Philosophy in Seventeenth-Century Rome*. Chicago: University of Chicago Press, forthcoming.

Stolzenberg, Daniel, ed. *The Great Art of Knowing: The Baroque Encyclopedia of Athanasius Kricher*. Stanford: Standord University Libraries, 2001.

Stone, Lawrence. *Family, Sex, and Marriage*. London: Weidenfeld & Nicholson, 1977.

Stone, Lawrence. "Literacy and Education in England, 1640–1900," *Past & Present* 42 (1969): 69–139.

Strauss, Gerald. *Luther's House of Learning: Indoctrination of the Young in the German Reformation*. Baltimore: The Johns Hopkins University Press, 1978.

Strauss, Gerald. "Success and Failure in the German Reformation," *Past and Present* 67 (1975): 59, 61–62.

Strauss, Walter L., ed. *The German Single-Leaf Woodcut, 1500–1550*. 4 vols. New York: Arbaris Books, 1975.

Strenski, Ivan. *Why Politics Can't Be Freed From Religion*. Oxford Wiley-Blackwell, 2010.

Struever, Nancy S. *The Language of History in the Renaissance: Rhetorical and Historical Consciousness in Florentine Humanism*. Princeton: Princeton University Press, 1970.

Stuckrad, Kocku von. *Locations of Knowledge in Medieval and Early Modern Europe: Esoteric Discourses and Western Identitites*. Leiden and New York: Brill, 2010.

—*Was ist Estoerik: Kleine Geschichte des geheimen Wissens*. Munich: Verlag C. H. Beck LHG, 2004.

Stubbe, Henry. *The Miraculous Conformist; An Account of Severall Marvaillous Cures Performed by a Stroaking of hands of Mr. Valentine Greatraik with a Physicall Discourse Thereupon. In a Letter to the Honourable Robert Boyle, Esq.* Oxford, 1666.

Styers, Randall. *Making Magic: Religion, Magic, & Science in the Modern World*. Oxford and New York: Oxford University Press, 2004.

Sullivan, Margaret. "Amazons and Aristocrats: The Function of Pyrocles' Amazon Role in Sidney's Revised *Arcadia*." *Playing with Gender: A Renaissance Pursuit*. Ed. Jean R. Brink, Maryanne C. Horowitz, and Allison P. Coudert. Urbana: University of Illinois Press, 62–84.

Swatos, William H. Jr. and Daniel V. A. Olson, ed. *The Secularization Debate*. New York: Rowman & Littlefield Publishers, Inc., 2000.

Swift, Jonathan. *Gulliver's Travels*. New York: W. W. Norton & Co, 2001 [1726].

Syfret, R. H. "Some Early Reactions to the Royal Society," *Notes and Records of the Royal Society* 7 (1950): 207–58.

Talbot, Charles W. "Baldung and the Female Nude." *Hans Baldung Grien: Prints and Drawings*. Ed. James H. Marrow and Alan Shestack. New Haven: Yale University Art Gallery, 1981, 19–37.

Tambiah, Stanley J. *Magic, Science, and Religion: The Scope of Rationality*. Cambridge: Cambridge University Press,1990.

Taufer, Alison. "The Only Good Amazon Is a Converted Amazon: The Woman Warrior and Christianity in the *Amadis Cycle*." *Playing with Gender: A Renaissance Pursuit*. Ed. Jean R. Brink, Maryanne C. Horowitz, and Allison P. Coudert. Urbana: University of Illinois Press, 35–52.

Taylor, Charles. *Sources of the Self: The Making of the Modern Identity*. Cambridge: Harvard University Press, 1992.

Temkin, Owsei. *The Falling Sickness: A History of Epilepsy from the Greeks to the Beginning of Modern Neurology*. 2nd ed. Baltimore: Johns Hopkins University Press, 1971.

Terrall, Mary. "Metaphysics, Mathematics, and the Gendering of Science in Eighteenth-Century France." *The Sciences in Enlightened Europe*. Ed. William Clark, Jan Golinski, and Simon Schaffer. Chicago, University of Chicago Press, 1999, 246–71.

Thomas, Keith. "An Anthropology of Religion and Magic, II." *Journal of Interdisciplinary History* 6 (1975): 91–109.

Thomas, Keith. "The Double Standard," *The Journal of the History of Ideas* 20 (1959): 195–216.

Thomas, Keith. *Man and the Natural World*. New York: Pantheon Books, 1983.

Thomas, Keith. "Puritans and Adultery: The Act of 1650 Reconsidered." *Puritans and Revolutionaries: Essays in Seventeenth Century History presented to Christopher Hill*. Ed. Keith Thomas and D. Pennington. Oxford: Clarendon Press, 1978.

Thomas, Keith. *Religion and the Decline of Magic. Studies in Popular Belief in Sixteenth and Seventeenth Century England*. London: Penguin, 1971.

Thorndike, Lynn. "Newness and Novelty in Seventeenth-Century Science and Medicine," *Journal of the History of Ideas* 12 (1952): 584–98.

Tiles, Mary. "Mathesis and the Masculine Birth of Time," *International Studies in the Philosophy of Science* 1 (1986): 16–35.

Todd, Dennis. "Three Characters in Hogarth's "Cuicularii"—and Some Implications," *Eighteenth-Century Studies* 16 (1982): 26–46.

Tomas, Lucia Tongiori. "Naturalistic Illustrations and Collections in Tuscany in the Eighteenth Century." *Science and the Visual Image in the Enlightenment*. Ed. William R. Shea. Science History Publications/USA, 2000, 111–36.

Trevor-Roper, Hugh. *The Crisis of the Seventeenth Century: Religion, the Reformation, and Social Change*. New York: Harper & Row, 1968.

Trevor-Roper, Hugh. *The European Witch-Craze of the Sixteenth and Seventeenth Centuries*. London: Penguin, 1969.

Trinkhaus, Charles. *In our Image and Likeness: Humanity and Divinity in Italian Humanist Thought*. 2 vols. Chicago: University of Chicago Press, 1992.

Troyansky, David G. *Old Age in the Old Regime: Image and Experience in Eighteenth-Century France*. Ithaca and London: Cornell University Press, 1989.

Turnbull, G. H. *Hartlib, Dury and Comenius*. Liverpool: University of Liverpool Press, 1947.

Tyndale, William. "Exposition of Matthew V, VI, VII." *The Works of William Tyndale*. Ed. G. E. Duffield. Philadelphia: Fortress Press, 1965.

Underdown, David. *Fire from Heaven. Life in an English Town in the Seventeenth Century*. London: HarperCollins, 1992.

Underdown, David. "The Taming of the Scold: The Enforcement of Patriarchal Authority in Early Modern England." *Order and Disorder in Early Modern England*. Ed. A. Fletcher and J. Stevenson. Cambridge: Cambridge University Press, 1985, 116–30.

Vanhaelen, Angela. *Comic Print and Theatre in Early Modern Amsterdam: Gender, Childhood and the City*. Aldershot: Ashgate, 2003.

Versluis, Arthur. "What is Esoteric? Methods in the Study of Western Esotericism." On-line journal *Esoterica* (2004) (http://www.esoteric.msu.edu/VolumeIV/Methods.htm).

Vickers, Brian. "Analogy and Identity." *Occult & Scientific Mentalities in the Renaissance*. Ed. Brian Vickers. Cambridge: Cambridge University Press, 1984.

Vickers, Brian. "Frances Yates and the Writing of History," *Journal of Modern History* 51 (1979): 287–316.

Vickers, Brian. "On the Function of Analogy in the Occult." *Hermeticism and the Renaissance*. Ed. Ingrid Merkel and Allen G. Debus. Washington: Folger Books, 1988, 265–92.

Vickers, Brian, ed. *Occult and Scientific Mentalities in the Renaissance*. Cambridge: Cambridge University Press, 1984.

Waddington, Raymond B. "The Bisexual Portrait of Francis I: Fontainbleau, Castiglione, and the Tone of Courtly Mythology." *Playing with Gender: A Renaissance Pursuit*. Ed. Jean R. Brink, Maryanne C. Horowitz, and Allison P. Coudert. Urbana: University of Illinois Press, 99–132.

Waddington, Raymond B. "The Sun at the Center: Structure as Meaning in Pico della Mirandola's *Heptaplus*," *The Journal of Medieval and Renaissance Studies* 3 (1973): 69–86.

Walker, D. P. *The Ancient Theology: Studies in Christian Platonism from the Fifteenth to the Eighteenth Centuries*. London: Duckworth, 1972.

Walker, D. P. "Francis Bacon and *Spiritus*." *Science, Medicine and Society in the Renaissance. Essays to honor Walter Pagel*. Ed. Allen G. Debus. New York, Science History, 1972, 121–30.

Walker, D. P. "Medical Spirits in Philosophy and Theology from Ficino to Newton." *Arts du spectacle et histoire des desidées. Receuiloffert en homage à Jean Jacquot*. Tours: Centres d'étudess superieures de la Renaissance, 1984.

Walker, D. P. *Spiritual and Demonic Magic from Ficino to Campanella*. London: University of London Press, 1958.

Walker, D. P. *Unclean Spirits: Possession and Exorcism in France and England in the late Sixteenth and Early Seventeenth Centuries*. London: Scolar, 1981.

Wallace, Anthony F. C. *Religion: An Anthropological View*. New York: Random House, 1966.

Wallis, Richard T. ed. *Neoplatonism and Gnosticism*. Albany: State University of New York Press, 1992.

Walsham, Alexandra. "The Reformation and "The Disenchantment of the World" Reassessed," *Historical Journal* 51 (2008): 497–528.

Walton, Michael T. "Boyle and Newton on the Transmutation of Water and Air, from the Root of Helmont's Tree," *Ambix* 27 (1980): 11–18.

Walzer, *The Revolution of the Saints: A Study in the Origins of Radical Politics*. London: Weidenfeld and Nicolson, 1966.

Warnicke, Retha. *Women of Renaissance and Reformation*. Westport and London: Greenwood Press, 1983.

Waswo, Richard. *Language and Meaning in the Renaissance*. Princeton: Princeton University Press, 1987.

Webster, Charles. *The Great Instauration: Science, Medicine and Reform, 1626–1660*. London: Duckworth, 1975.

Webster, Charles, ed. *Health, Medicine and Mortality in the Sixteenth Century*. Cambridge: Cambridge University Press, 1979.

Werblowsky, R. J. Zwi."Messianism in Jewish History." *Essential Papers on Messianic Movements and Personalities in Jewish History*, ed. M. Saperstein. New York: New York University Press, 1992.

Westfall, Richard. "Newton and Alchemy." *Occult and Scientific Mentalities in the Renaissance*. Ed. Brian Vickers. Cambridge: Cambridge University Press, 1984.

Westfall, Richard. "Newton and the Hermetic Tradition." *Science, Medicine, and Society in the Renaissance*. Ed. Allen G. Debus. New York: Science History Publications, 1972.

Westfall, Richard. "The Rise of Science and the Decline of Orthodox Christianity: A Study of Kepler, Descartes and Newton." *God & Nature: Historical Essays on the Encounter between Christianity and Science*. Ed. David C. Lindberg and Berkeley: University of California Press, 1986, 218–37.

Westfall, Richard. "The Scientific Revolution Reasserted," *Rethinking the Scientific Revolution*. Ed. Margaret J. Osler. Cambridge: Cambridge University Press, 2000, 41–58.

Westman, Robert S. "The Astronomer's Role in the Sixteenth Century: A preliminary Survey," *History of Science* 18 (1980): 105–47.

Wiesner, Merry E. *Women and Gender in Early Modern Europe*. Cambridge: Cambridge University Press, 1993.

Wiesner-Hanks, Merry E. *Christianity and Sexuality in the Early Modern World: Regulating Desire, Reforming, Practice*. London and New York: Routledge, 2000.

Wilkins, John. *Essay Towards a Real Character and a Philosophic Language*. London, 1668,

Williamson, Arthur H. *Apocalypse Then: Prophecy and the Making of the Modern World*. Westport, Conn. and London: Praeger, 2008.

Williamson, George. "Mutability, Decay, and Seventeenth Century Melancholy," *ELH* 11 (1935): 121–51.

Wilson, D. C. "Contraries in Sixteenth Century Scientific Writing in France." *Essays presented to* C. M. *Girdlestone*. Ed. E. T. Dubois et al. Newcastle upon Tyne, 1960, 195–217.

Wilson, Stephen. *The Magical Universe: Everyday Ritual and Magic in Pre-Modern Europe*. London and New York: Hambledon and London, 2000.

Wiltenburg, Joy. *Disorderly Women and Female Power in the Street Literature of Early Modern Germany and England*. Charlottesville: University of Virginia Press, 1992.

Wind, Edgar. *Pagan Mysteries in the Renaissance*. New York: W. W. Norton, 1968.

Winner, Anna K. *The Basic Ideas of Occult Wisdom*. Wheaton: The Theosophical Publishing House 1970.

Wittkower, Rudolph. "Marvels of the East," *Journal of the Warburg and Courtauld Institutes* 5 (1942): 159–97.

Wittkower, Rudolph and Margot. *Born Under Saturn: The Character and Conduct of Artists*. New York: W. W. Norton and Company, 1969.

Wokler, Robert. "The Enlightenment Project and its Critics," *Poznan Studies of the Sciences and the Humanities* 58 (1997): 13–30.

Wood, Anthony. *The Life and Times of Anthony Wood*. Ed. Andrew Clark. Oxford Historical Society. 5 vols. Oxford Clarendon Press, 1891–1900.

Woodhouse, J. R. *Baldesar Catiglione*. Edinburgh: Edinburgh University Press, 1978.

Woods, Susanne. "Amazonian Tyranny: Spenser's Radigund and Diachronic Mimesis." *Playing with Gender: A Renaissance Pursuit*. Ed. Jean R. Brink, Maryanne C. Horowitz, and Allison P. Coudert. Urbana: University of Illinois Press, 52–61.

Yates, Frances A. *The Art of Memory*. London: Routledge & Kegan Paul, 1966.

Yates, Frances A. *The French Academies of the Sixteenth Century*. New York: Routledge & Kegan Paul, 1975.

Yates, Frances A. *Giordano Bruno and The Hermetic Tradition*. London: Routledge & Kegan Paul, 1964.

Yates, Frances A. "The Hermetic Tradition in Renaissance Science," *Art, Science, and History in the Renaissance*. Ed. Charles S. Singleton. Baltimore: Johns Hopkins Press, 1968.

Yates, Frances A. *The Rosicrucian Enlightenment*. London: Routledge & Kegan Paul, 1972.

Zika, Charles. "Baldung [Grien], Hans (1484–1545)." *Encyclopedia of Witchcraft: The Western Tradition*. Ed. Richard M. Godlen. Santa Barbara: ABC-CLIO, 2006. Vol. 1, 80–82.

Zika, Charles. "Dürer's Witch, Riding Women and Moral Order," *Dürer and his Culture*. Ed. Dagmar Eichberger and Charles Zika. Cambridge: Cambridge University Press, 1998, 118–42 [republished in *Exorcising Our Demons*, ch. 8].

Zika, Charles. *Exorcising our Demons: Magic, Witchcraft and Visual Culture in Early Modern Europe*. Leiden: Brill, 2003.

Index

About the Author

ALLISON P. COUDERT is the Paul and Marie Castelfanco Chair in the History of Religion in the Religious Studies Program at UC Davis. She grew up in New York City, and received her BA from Vassar College and PhD from The Warburg Institute, University of London. Her primary focus of interest is in the interrelation between Religion and Science as well as the way both have historically shaped ideas about gender. She has also written extensively on the historical relations between Christians and Jews and the role that the mystical teachings of the Jewish Kabbalah played in the emergence of modern science (*The Impact of the Kabbalah in the Seventeenth Century*; 1998; *Leibniz and the Kabbalah*, 1995). Her most recent articles include, "Sewers, Cesspools, and Privies: Waste as Reality and Metaphor in Pre-modern European Cities," *Urban Space in the Middle Ages and the Early Modern Age* (2009) and "Faith and Religion," *A Cultural History of Childhood and Family*, vol. 4 (2010).

Praeger Series on the Early Modern World

Apocalypse Then: Prophecy and the Making of the Modern World
Arthur H. Williamson

Why Machiavelli Matters: A Guide to Citizenship in a Democracy
John Bernard

Holy War and Human Bondage: Tales of Christian-Muslim Slavery in the Early-Modern Mediterranean
Robert C. Davis